INVESTING IN RURAL EXTENSION: STRATEGIES AND GOALS

Edited by

GWYN E. JONES

Agricultural Extension and Rural Development Centre,
University of Reading, UK

ELSEVIER APPLIED SCIENCE PUBLISHERS
LONDON and NEW YORK

ELSEVIER APPLIED SCIENCE PUBLISHERS LTD
Crown House, Linton Road, Barking, Essex IG11 8JU, England

Sole Distributor in the USA and Canada
ELSEVIER SCIENCE PUBLISHING CO., INC.
52 Vanderbilt Avenue, New York, NY 10017, USA

WITH 10 TABLES AND 17 ILLUSTRATIONS

© ELSEVIER APPLIED SCIENCE PUBLISHERS LTD 1986

British Library Cataloguing in Publication Data

Investing in rural extension: strategies
  and goals
  1. Agricultural extension work
  I. Jones, Gwyn E.
  630.7'15   S544

Library of Congress Cataloging-in-Publication Data

Investing in rural extension.

  Based on a conference held at the University of Reading,
Sept. 1985, and organized by the Agricultural Extension and
Rural Development Centre.
    Bibliography: p.
    Includes index.
    1. Agricultural extension work — Congresses.
  2. Rural development — Congresses.   I. Jones, Gwyn Evans.
  II. University of Reading, Agricultural Extension and
  Rural Development Centre.
  S544.I59   1986   630'.7'15   86-13371

  ISBN 1-85166-020-8

The selection and presentation of material and the opinions expressed in this publication
are the sole responsibility of the authors concerned.

Special regulations for readers in the USA

This publication has been registered with the Copyright Clearance Center Inc. (CCC),
Salem, Massachusetts. Information can be obtained from the CCC about conditions
under which photocopies of parts of this publication may be made in the USA. All other
copyright questions, including photocopying outside of the USA, should be referred
to the publisher.

Phototypesetting by Tech-Set, Gateshead, Tyne & Wear.
Printed in Great Britain by Galliard (Printers) Ltd, Great Yarmouth.

Preface

This volume is the outcome of a conference, held at the University of Reading in September 1985, which was organized by the staff of the University's Agricultural Extension and Rural Development Centre.

Over the past two decades, the AERDC has become one of the main international centres offering postgraduate courses to those engaged in agricultural and other forms of rural extension, and in agricultural education. In addition, its staff members have been active in research work and as consultants, and in providing short courses overseas and in Britain. The experience gained in the association with students from very many parts of the world and in the varied international activities has produced a critical appreciation of the many roles which extension workers can play among rural people and of the difficulties which their work faces. In considering how best to celebrate the twentieth anniversary of the AERDC it seemed appropriate that it should arrange and host an international conference. This would aim to focus on a theme which now dominates much of the discussion surrounding rural extension, namely, the utility of the investments being devoted to the work in relation to its strategies and its goals.

The organization of the conference was only possible as a result of much generous support and many forms of assistance received. The Organizing Committee wishes to acknowledge this support, financial and otherwise, given by: the W. K. Kellogg Foundation; the World Bank; the British Council; the Agricultural Development and Advisory Service (ADAS) of the Ministry of Agriculture, Fisheries and Food; the British Agricultural Export Council; the Plant Protection Division of Imperial Chemical Industries PLC; FBC Ltd; Dalgety PLC; Barclays

Bank PLC; ULG Consultants Ltd; and Hunting Technical Services Ltd. The occasion was honoured by the Rt Hon. Timothy Raison, M.P., the Minister of Overseas Development, who kindly agreed to open the week's proceedings.

In my capacity as Chairman of the Organizing Committee, I would wish to acknowledge the guidance and continuous support provided by my colleagues who served on the committee: Maurice Rolls, Peter Oakley, Chris Garforth, Patricia Goldey and John Best. I am also grateful to those who acted as rapporteurs for the discussion groups who, apart from Bob Kern, were all AERDC staff members (John Best, Chris Garforth, Patricia Goldey, Howard Jones, Peter Oakley, Charles Stutley and Bob Wake), and to the various participants who agreed to chair the plenary and discussion group sessions. The Centre's secretarial staff willingly provided all kinds of essential assistance, and I am especially indebted to Marilyn Seabrooke who helped with the detailed arrangements and in innumerable other ways. Finally, I would wish to express my gratitude to Maurice Rolls who, as Director of the Centre and as always, gave his constant encouragement and much sage advice.

GWYN E. JONES

Contents

Policy, and the Planning of Extension

Extension Strategies and Approaches

Extension Methods

Resources and Management

Research, Monitoring and Evaluation

List of Contributors

HARTMUT ALBRECHT
Professor of Agricultural Extension, Institute for Rural Sociology, Agricultural Extension and Applied Psychology, Universität Hohenheim, Postfach 700562, D 7000 Stuttgart 70, Federal Republic of Germany.

ROBERT BRUCE
Professor of Extension Education, Department of Education, Cornell University, Ithaca, New York 14850, USA.

A. H. BUNTING
formerly Professor of Agricultural Development Overseas, Faculty of Agriculture and Food, University of Reading, Earley Gate, Reading RG6 2AT, UK.

WILLIAM CHANG
Chief Extension Officer, Department of Agriculture (Headquarters), Kuching, Sarawak, East Malaysia.

G. CAMERON CLARK
Consultant on FAO's Small Farmers Development Programme in South Asia, 98 Country Club Drive, Kingston, Ontario, K7M 7B6, Canada.

E. DEXTER
Head, ADAS Extension Development Unit, Ministry of Agriculture, Fisheries and Food, Government Buildings, Coley Park, Reading, RG1 6DT, UK.

S. P. DHUA
Chairman and Managing Director, Hindustan Insecticides Ltd, Hans Bhawan (Wing 1), Bahadur Shah Zafar Marg, New Delhi 110 002, India.

ROBERT E. EVENSON
Professor of Economics, Economic Growth Centre, Department of Economics, Yale University, PO Box 1987, Yale Station, New Haven, Connecticut 06520-1987, USA.

GERSHON FEDER
Economist on Staff of The World Bank, 1818 H Street NW, Washington, DC 20433, USA.

JOHN FYFE
Overseas Manpower and Employment Adviser, Overseas Development Administration, Eland House, Stag Place, London, SW1E 5DH, UK.

CHRIS GARFORTH
Lecturer, Agricultural Extension and Rural Development Centre, University of Reading, Earley Gate, Reading RG1 5AQ, UK.

PETER W. C. HOARE
Extension Training Officer, Highland Agricultural and Social Development Project, 40/20 Soi 3, Huai Khaed Road, T. Suthep, Chiangmai 50000, Thailand.

JOHN HOWELL
Deputy Director, Overseas Development Institute, 10–11 Percy Street, London, W1P 0JB, UK.

PAUL INGRAM
Senior Agricultural Officer, ADAS Headquarters, Ministry of Agriculture, Fisheries and Food, Great Westminster House, Horseferry Road, London, SW1P 2AE, UK.

GWYN E. JONES
Senior Lecturer, Agricultural Extension and Rural Development Centre, University of Reading, London Road, Reading, RG1 5AQ, UK.

PETER OAKLEY
Lecturer, Agricultural Extension and Rural Development Centre, University of Reading, London Road, Reading, RG1 5AQ, UK.

MALCOLM J. ODELL JR
Managing Partner, Synergy International, 418 Main Street, Amesbury, Massachusetts 01913, USA

JOAN ALLEN PETERS
Senior Lecturer in Home Economics, Bath College of Higher Education, 11 Somerset Place, Bath, BA1 5SJ, UK.

CONSTANTINOS L. PHOCAS
Director of Agriculture, Ministry of Agriculture and Natural Resources, Nicosia, Cyprus.

DONALD C. PICKERING
Assistant Director, Agriculture and Rural Development Department, The World Bank, 1818 H Street NW, Washington, DC 20433, USA.

NIELS RÖLING
Professor of Extension Education, Department of Extension Education, Agricultural University, Hollandsweg 1, 6706 KN Wageningen, The Netherlands.

MAURICE J. ROLLS
Director, Agricultural Extension and Rural Development Centre, University of Reading, London Road, Reading, RG1 5AQ, UK.

JOHN F. A. RUSSELL
formerly Rainfed Crops Adviser, Agriculture and Rural Development Department, The World Bank, Resident Staff in Indonesia, PO Box 324/ JKT, Jakarta 12940, Indonesia.

ROGER H. SLADE
Senior Economist, South Asia Projects Department, The World Bank, 1818 H Street NW, Washington, DC 20433, USA.

ANNE W. VAN DEN BAN
Consultant and formerly Professor of Extension Education, Agricultural University, Foulkesweg 82a, 6703 BX Wageningen, The Netherlands.

STEVE WIGGINS
Lecturer, Department of Agricultural Economics and Management, University of Reading, Earley Gate, Reading, RG6 2AR, UK.

Introduction

The need for development in rural areas is one of the dominant issues of concern in today's world. It is only relatively recently, however, that the crucial importance of this has become widely recognized and accepted. Rural development involves the acceptance by rural people of the possible, often new ways and means of developing their economies; usually this implies the development of their agriculture (defined in its broadest sense to include all uses of the land and many directly associated productive activities). But, this is largely a means to a greater end. Rural development also embraces an active concern for improvement in the welfare and well-being of all rural inhabitants who, even in times free from crises (such as famines or other disasters), include a high proportion of the world's poorest people. Among the critical issues, therefore, are how to ensure that the basic needs of life are available to rural people, and how to enable them to gain at least adequate livelihoods by means which are equitable and sustainable, to enjoy ways of living which are culturally appropriate and satisfying, and to assure that these have a capacity for further improvement.

A large variety of factors are involved in rural development which are both complex in themselves as well as in their interrelationships. These can be (and in recent years have been) analysed and discussed from many disciplinary and ideological perspectives, and at different theoretical and conceptual levels. Many forms of policy and policy implementation which aim to ameliorate rural conditions have been devised and tried by authorities ranging from international agencies and national governments to local communities initiating their own activities. In practice, however, whatever view is held, and whatever kind

1

of intervention or action is attempted, much of the development which is possible rests on the rural people themselves, as communities, families and individuals. Given that their basic requirements for survival are satisfied, they then need to be enabled to improve their living conditions in ways which are compatible with their own ideals, their culture, and their environment.

One of the important (though often neglected) ways by which rural people can be assisted in this is through extension work. Traditionally, the activities of extension workers have been largely confined to agriculture, but by now comparable approaches have become adopted increasingly to assist people with many other facets of their lives. In essence, extension work involves a collaborative relationship between rural people and extension workers (and thereby the agencies to which they belong) in a deliberate process. Ideally, the aim is to offer relevant, reliable, and usable information, advice and guidance to the people on acceptable technological, socio-economic and other ways of overcoming their problems or of realizing betterment in their lives. However, in reality, one or both of two problems commonly recur: the necessary information is not known and is thus a matter which requires research or investigation; and the strategies, methods, and relationships by which information and advice is conveyed to the people are insufficient, or not sufficiently effective, to persuade them of its efficacy. Because it is difficult to evaluate the effects of extension activities convincingly, their content, as well as the ways in which they are organized and managed, are thus often regarded as open to criticism. Commonly, if development occurs in a rural area which is regarded as 'successful', the praise is given to other, more tangible or easily quantifiable factors (e.g. the adoption of better farming practices or systems, or innovations emanating from research centres, or infrastructural improvements); if there is a 'failure', the inadequacies of extension work are likely to be blamed.

Yet, in the rural areas of most developed countries as well as in those of the Third World, extension is an activity in which international agencies, governments, non-governmental organizations, and, not least, the rural people are making substantial investments. Whatever guiding principles underlie it, by its nature extension work is costly in social as well as economic terms. It has become one of the main resources which are being managed in order to support and assist the development of other rural resources — the people and their natural environment — and their capacities. Many of the results which can properly be attributed to it, however, even if only in part, are often slow to materialize. Much recent debate has been focusing on the value of the extension activities which are being conducted among farmers, their families, and other rural people. Increasingly, therefore, the ideas, policies and broad strategies which govern rural extension work, as well as its organization, management, and methods of operation, are under scrutiny.

This volume is intended as a contribution towards furthering and enhancing this debate. It consists of the proceedings of an international conference which was held at the University of Reading in September, 1985. This was conceived by the staff of the University's Agricultural Extension and Rural Development Centre to celebrate the twentieth anniversary of the establishment of the AERDC. The broad theme chosen for the conference, which forms the title of the volume, was intended to reflect and draw attention to the importance of considering rural extension as a developmental process which involves intellectual, social, and economic investments.

The structure of this volume corresponds in large measure to that of the conference. The first part is concerned with the dimensions of rural extension work. Initially, this is considered in a paper written by three staff members of the AERDC to provide a general overview of the theme with three prepared comments upon it and arising from it. This is followed by the three papers which formed the plenary lectures at the conference. The following sections present the contributions which were invited to open the deliberations in five areas under which simultaneous discussion groups were arranged during three days of the conference. These have been rearranged under the five topics, with brief summaries of the discussions which were enriched by the experiences of participants and often by constructive controversy. Inevitably, there is some overlap between papers and discussion in different sections, but this reflects the interconnectedness of the topics and the progressive nature of the discussions during the conference. The papers, being intended to open discussion, frequently raise questions rather than provide answers. The discussions indicate not only the range of issues arising from the lead papers but also that many questions remain unresolved. In total, the proceedings show the importance of recognizing the many complexities which are involved in considering the value of extension work in today's world, and in the future, as a means of helping rural people to attain sustainable livelihoods and satisfying lives.

1

The Dimensions of Rural Extension

MAURICE J. ROLLS, GWYN E. JONES and CHRIS GARFORTH
Agricultural Extension and Rural Development Centre, University of Reading, UK

INTRODUCTION

This paper is intended to set the general scene on 'Investing in Rural Extension' by allowing some common assumptions and precepts to be discussed which might otherwise pass unstated. It is also a prologue to discussion and enlargement by three discussants, and to three theme papers which will each address a fundamental way in which extension can be conceived. It neither defines in the narrow terminological sense nor reports on major new work that leads to some reformulation of basic issues: to define adequately the complex and varied ideas that are encompassed by the term 'extension', as this relates to development of rural areas, and to say something new about them, would be daunting tasks.

ORIGINS OF EXTENSION

The mid nineteenth century saw the start of institutionalization and public funding of organized extension activity concerned with agriculture and rural life. But for centuries before this began to occur there had existed a very informal sharing among rural people of ideas, beliefs, knowledge and information in the form of both common and more unusual experience of solving problems in their farming and everyday life.

In Europe, extension work began with the creation of a small force of itinerant 'agricultural instructors' in Ireland between 1847 and 1851, at the time of the serious potato crop failures and the famine. As a response to the crop failures, the fourth Earl of Clarendon (the Lord Lieutenant of Ireland at the time) sought to stimulate changes in the cropping and the associated husbandry practices of impoverished Irish small farmers with the aims of reducing the dependence of the peasantry on potatoes and creating a system of farming that was much less prone to attack by the potato blight fungus. This was done not by using market forces or legislative authority, but by means of activities which were essentially informative and educational, organized so that reliable innovations became quickly available to large numbers of very small farmers who were in a situation of crisis. The impact was substantial and beneficial. Similar experiences also, to a large extent, in response to crises facing small farmers were soon gathered in Germany and France in the 1860s and 1870s, where early instructional and advisory work among peasant farmers was initiated. Before the beginning of this century, advisory or extension services for farmers, often including in their work large elements of farmer instruction in better husbandry methods and skills, had been established in most European countries, North America, and many tropical parts of the world. In many cases the latter were strongly influenced by colonial authorities to develop new farming systems to produce export crops to be processed in Europe. Extension services, often with this objective, were operating (for example) in the Indian subcontinent by the 1860s, in the Caribbean by the late 1880s, in Egypt by the 1890s, and in various African countries by the first decades of this century.

Looked at from our interests today, the European experiences were an impressive demonstration of how investment in farming in the form of relatively underused research results could be used together with land, labour and the more tangible forms of capital for farming inputs to bring about an improvement in individual lives, and thus in rural society as a whole. The application of scientific investigation to agriculture and the institutionalization of agricultural research had been developing in Britain, in continental Europe, and in North America throughout the nineteenth century, ahead of the initiation of formal extension work and agricultural education. This led to the development of a concept of extension work as an integral part of, or at least a necessary adjunct to, a knowledge generating (research) service which spans the identification of needs and problems, the generation and testing of new or improved technologies, and the dissemination of relevant, reliable information (even firm recommendations) to the potential users. It produced the idea of a conceptually simple system, with components of research, extension and education which evolved during the first half of this century and which served for some time in the 1950s and 1960s as a framework for explaining and intervening in processes of agricultural change and development, most notably in the United States where the components were concentrated on the universities.

THE DEVELOPMENT OF EXTENSION SERVICES

Probably the most important developments in this research and technology-based view of extension, in the last thirty years, have been in four directions. Arguably, the first comes from the development projects conducted by Shell in the 1950s and 1960s, especially by Borgo a Mozzano in Italy, which attempted to formalize the process of extension work in a community into recognizable and sequenced phases. More importantly, it also sought to explore the relationship between the costs of employing an agriculturist as an agent of change and the return, at least in the short-term, in terms of additional outputs and higher farm incomes which accrue to the community. Impetus was thus given to the second direction of the development of extension, namely, the idea that it is a public service which can be, at least to some extent, managed to give optimum productivity on an investment of public funds. Perhaps the most comprehensive attempt at management is the recent (since 1977) introduction of the 'Training and Visit' system of extension in which objectives, accountability and regular schedules of activities are strongly emphasized. The third direction of development has been in methodologies and methods, where many important advances have been made in understanding the communication processes involved in extension work and especially how improvements in information technology can be used to complement the personal approach.

Finally both the concept and the practice of extension have increasingly been applied in areas of rural life outside the strict confines of agriculture. Institutionally, this development has sometimes taken place within agricultural extension agencies: under the name, for example, of Home Economics. But the principles and methods of extension work are now being deployed in diverse institutions, both within government departments and in non-governmental organizations. This should cause no surprise. Examples of the value of knowledge in crisis situations, analogous to those already noted in agriculture, are sadly common in health and nutrition problems which especially affect the old, young and disadvantaged people in society. Relatively simple preventive and curative measures are sometimes available, though unknown and unused until a health worker or nutritionist is deployed among the communities concerned. Similarly, social innovations such as cooperatives, credit groups or marketing associations often depend for their success on effective communication within the agencies and between these and their members, and on the level of understanding of what is involved in their effective management. Extension services, either general or more specialized, exist in many countries to provide information, advice and education relating to many facets of rural life and its improvement, and in some cases extend to small businesses and other forms of income-earning activities open to the rural population from non-farming sources. We have thus reached the idea of 'rural' rather than more narrowly 'agricultural' extension.

What started as itinerant instruction, with a heavy dependence on individual enthusiasm to 'do good' and a somewhat esoteric interpretation, has become more and more a managed activity arising from policy decisions with an emphasis on planning and deciding how to implement and operate the work, aided by assessing and evaluating the outcomes, leading to modified, adapted and new forms of work. The resulting patterns of extension vary considerably within and between countries.

The role and practice of extension, which now rarely starts in an entirely fresh way, has thus evolved through historical development and adjustment to the varied and changing internal conditions in a country or area and to external pressures. Of special importance in this have been the values, norms, knowledge, ideas, and aspirations of specific societies. Thus, it is when the range of expectations which exist — in society about extension services, and in extension organizations about their staff, and of rural people about extension workers — is appreciated that simplicity and uniformity must be abandoned.

We have now encountered some of the often unstated assumptions alluded to at the beginning of this paper, which are discussed more explicitly below. The first of these is that the provision of new technical information in itself enables rural people to improve their productivity, health or other aspects of their lives; the second, that technological innovations should properly originate from institutionalized scientific research; the third, that knowledge created by research and propagated by extension services is necessarily superior to the knowledge already available to and used by rural people; and finally, that the calculus of improvement in the lives of individual members of society inevitably produces an overall net gain in social well-being.

DIMENSIONS OF EXTENSION

At its most basic, we are concerned with social and economic change. This has both absolute ends, in increasing national wealth and well-being, and relative goals in terms of the distribution of wealth, equity, and equality of social rank within a nation. Farmers and farms are extremely varied in all their characteristics in the agricultural industries of all countries. It is necessary, therefore, to question not only how extension can best serve the industry, and *all* sectors of it, but also how the research services function to generate the information and innovations which are their main goals, and the implicit role of extension and related services to disseminate. In particular, one can ask by whom is the research policy determined and how well are its programmes and their results matched to the needs of farmers and their families? Considerable changes have occurred in the approach to answering these questions in the past twenty years.

Put simply, there has been a substantial erosion of the simple belief that the research services invariably and inevitably produce technically sound and economically profitable possibilities for change in production systems. Doubts have also grown that the incorporation of research findings into the national knowledge system, and their acceptance by rural people, are both retarded by inadequacies in the extension services and resistances by rural communities. In agriculture, experience has shown that innovations developed by research, or imported from other countries, or transferred from other farming systems, are not always an unequivocal success. Some have eventually been shown to be less productive, or inherently more risky, than existing practices. Others have enhanced the productivity and economic well-being of some farm families while leading relatively to the further impoverishment of others. Some have had unintended environmental consequences which create long-term problems for the industry and society in general. Existing production systems have often been developed over many generations of trial and error to form complex farming systems which make good use of the local conditions and available resources and which give some insurance against the risks inherent in all farming. Sometimes, single changes developed from specific research projects can be introduced relatively easily into such a system. More typically, however, there are packages of changes and adjustments to make which call for managerial skills, new knowledge and other resources that may simply not be available. It may be that research staff do not always know enough about the intricacies and complexities of existing production systems to be able to specify the appropriate research problems. Even when they are, there is need for sensitivity to the community's customs and traditions and its indigenous knowledge system, and the complex ways in which these have to assimilate, and adjust to, a more modern agricultural information system with its different values and practices which are derived from research and innovation. Two further points can be made. First, research work may need to adapt its procedures and its understanding of farming systems to indigenous production systems. Secondly, farmers are a good and often valid source of innovations that can be verified, refined and disseminated by the research and extension service. The growing emphasis on farming systems research and our increasing understanding of (and respect for) the existing farming practices in many 'less developed' countries are positive gains from a re-assessment of policies and practices of agricultural research. Similarly, the idea of adaptive research has evolved as a means to avoid the dissemination of generalized recommendations which may be based on the work of distant research stations. Just as forms of agricultural literature (or 'media') pick up audiences that can use them, so research will attract its particular adopters: if these turn out to be only a minority of the total population, one has to ask what is being done for the majority?

A CAUTION

It is perhaps time to put ourselves in our place. We must follow convention in remembering that extension and knowledge, as we have so far discussed it, are only part of the factors promoting change that include supplies of inputs and marketing outlets for production at higher technological and productivity levels. Extension services which are closely linked to research services and which see this linkage to be their essential nature are thus part of an approach to change. It is essentially one of pressure for innovation brought to bear on communities by policies and institutions whose base, even when democratic and humane and in the best national (and individual) interests, is inevitably distanced geographically from most of the rural communities themselves. The discussion now turns to whether this is entirely satisfactory, and what other conceptions of extension are possible.

NEW DIMENSIONS

One such alternative is to retain the identification and dissemination of useful information as the central task of extension, but to focus attention on rural society rather than outside agencies as the initiator of the process. This change of perspective allows us to see the authority for intervention in the social and economic affairs of rural society as residing within that society itself rather than with policy makers and administrators. Also, we can emphasize the invention, adaptation and management of technology by people in the community rather than by staff of national and international research institutes. We can examine, too, a much richer set of communication processes than the creation and transmission of 'messages' by extension agencies. We can explore the capacity among rural people not only to receive but actively to seek out information, knowledge and skills; the sharing of ideas and experience that goes on within and between communities; and the making known of rural views and ideas to researchers, policy makers and urban society.

If this perspective is adopted, the role of research and extension is no longer to produce and disseminate defined solutions to problems that exist, judged by a gap between what is actually done and what better technology could potentially achieve. Rather, it is assistance to the community (through whatever associations exist) in understanding itself and its present use of its resources, and in analysing for itself the options and directions open to it in achieving progress towards satisfying its own needs and goals. This would mean, in effect, an exchange — knowledge about the existing farming systems and ways of life being given by the people to the invited agricultural and social scientists, in return for ways of thinking and analysing contributed by the scientists and advisers to the people. The implications of such a process in terms of resource use, time needed, care over decision-making, role clarification and compatibility between local, regional and national development, are possible major obstacles to its adoption.

There are further implications of this alternative view, beyond a change in the relationship between scientist and rural people. For example, a strategy of inducing change gives way to one of expanding the range of choices open to rural families and communities. The former implies the predetermination of objectives, followed by methods of a more or less persuasive nature to bring about the desired changes which, in agriculture, are often described as changes of farmer behaviour. The idea of choice, however, implies that extension should be a matter of giving information and advice in a situation where the decision-making remains with the recipients of that advice. Strategies of change or persuasion are widespread, and are at the heart, for example, of many agricultural extension systems based on the 'Training and Visit' model. But strategies of choice, also, have a long history and are implicit in the common approach to agricultural extension in north western Europe.

In practice, change and choice may not be as separate and distinct as we have stated, in that the ultimate decision on whether or not to follow a particular course of action remains with the individual, household or community. There are also further options through strategies of enabling (through education and training) and giving protection to those disadvantaged by change (through welfare). However, the idea of providing a means whereby rural people can clarify their own perceptions of their situation, and receive the information needed to generate the options available to them, whilst leaving them open to select the action they consider best suited to their purposes, is substantially the way in which some of the most effective agricultural advisory services have evolved. Lack of sufficiently highly trained field staff (and of other related resources), and the relatively large number and small size of farm units, are among the reasons why the change strategy has been more commonly adopted within agricultural extension in less developed countries.

Another implication concerns the use of communication channels, which is in part a consequence of the development and growth of educational aids and communications technology. Whilst making the media potentially more persuasive (and possibly manipulative), the uses to which some kinds of educational technology can be put to disseminate information of relevance to rural people are not only in the direction of strengthening the impact of the messages brought to a community. They can, at least in theory, provide the means for the individuals in a community both to make their views and ideas known to those in positions of influence in government and other institutions, and also to share their experiences and ideas with other rural families and communities. Communications technology also gives people much greater potential control over their access and exposure to information, enabling them to take the initiative as information seekers rather than adopt a passive role as information recipients.

A still more radical view of extension emerges when the realities of the rural society within which information is generated, made available and put to use are brought more sharply into focus. It can be argued that the heterogeneity of rural society severely limits the impact of a research/technology/dissemination approach to the promotion of rural development by imposing constraints on access to, and use of, available information. This heterogeneity refers not only to the differences in scale (and therefore technology) between farms, which make certain agricultural innovations appropriate to some but not to others. It implies also differences in fundamental interests — between land-owners and tenants, farmers and landless families, men and women, crop growers and livestock owners, rich and poor. And beyond these differences lie imbalances in power and influence (over resources and over other people) which lead to biases in the identification of problems to be researched as well as in the access of different members of society to the benefits of research. The role of extension, on this view, expands to include the enabling of those without power and influence to benefit from developments in productive opportunities and to resist changes that will make them worse off than before.

This view, also, has implications for ways in which extension workers interact with rural communities. There is an explicit recognition of different groups, rather than an assumption of uniformity of interests. Analyses of problems and potential, the drawing up of options, the search for suitable technology become processes to be carried out with specific groups rather than with 'the community'. Communications technology serves a particular function: it is used to help groups to analyse the situations and constraints that confront them, and also where necessary, to lobby those in positions of influence. Further, there is an emphasis on supporting the development of self-sustaining groups as a countervailing force in rural society, through which powerful individuals, groups and institutions can be confronted.

What seems to be happening within and beyond the conventional approaches to extension (conceived both as an acceptable, proper and legitimate way for society to aim to achieve certain aspirations and as a simultaneous reflection of its social and economic values), is a convergence of pragmatism (coming from experience of development work) and development theories. The focus of such convergence is on understanding existing social and agricultural systems before intervention is attempted, and an involvement and interaction with these systems by producers, research workers and extension staff when technology is changed at village level.

Naturally, these steadily expanding views of the extension function in development have been accompanied by questioning and criticism of its role and value and what can realistically be expected. To some extent the questioning is much concerned with what a society (and its government) wants of its rural resources (the land and the skills and abilities of the rural people), and for the benefit of whom, as these are expressed through the leadership, the market and the institutions of society (such as the media which influence and express public opinion). It is also concerned with questioning the role of its extension services. The forms taken by the criticisms of extension work, however, have first been to highlight certain inadequacies at an operational level. Thus, it has been judged variously, sometimes to be too narrow and unambitious, and at other times to be trying to do too much with inadequate resources and staff. Its orientation is sometimes said to be too technological or inappropriate. There is widespread comment of the inadequate quality of policy direction and management in the more traditional extension systems. Second, fundamental critiques have been offered about how new technology is generated, by whom, and for what purposes, and how the extension services should function at the community level.

Of greater importance are the underlying issues of whether the extension processes conceived by the institution, and its organization and its personnel (assuming these can be made to be identical or at least consistent) are to be conducted in a coercive or in a collaborative and participatory style, and with what degree of planning and rigidity. This is close to questioning the degree of autonomy to be given or allowed to extension services and their staff, for example, in having a separate identity rather than close integration with research services or other policies or with administrative ministries, or with the rural communities themselves. There are also functional questions about the relationships of extension to national policies and whether the concern is to be for agriculture (crops and livestock, or separate sectors; commercial and subsistence crops, or separate), or for agriculture including a variety of land uses such as forestry and grazings, or for the ecological concerns about land such as soil and wildlife conservation and protection of the environment. These qualitative aspects of agricultural production go further, into concerns about animal welfare, the use of chemicals, and the exploitation of labour. In both industrialized and less developed countries, agricultural policy is inevitably concerned with socio-economic conditions in the rural areas, such as land distribution and tenure, land inheritance and succession, rural employment and job opportunities, income generation from non-farm activities, rural services and amenities. Faced with these possible dimensions of extension, the relief with which a system of simple messages delivered through simple lines of control (such as in the Training and Visit system) may be greeted, is entirely understandable. The training requirements of extension staff in an open-ended, entirely individual client oriented service are certainly highly complicated, and perhaps unattainable.

These matters are variously resolved by countries. The resulting objectives with their relationships to policy and the legitimate authority which initiates it, and to the extension activities and their outcomes which contribute in part to further policy change and action (Fig. 1) become an issue of resource allocation and an assessment of the costs and benefits of extension. At this point we reach the question of investment in rural extension.

INVESTING IN EXTENSION

The outcome of the investment, as it has been discussed here, is the provision of *public information* about agriculture, needed to formulate and understand what is involved in policy for a vital industry, and the dissemination of *working knowledge*, needed by producers and those in the agricultural support services if they are to be effective in using the technology immediately available and what is constantly being developed as innovative possibilities. Some of this information and knowledge will be self-generated and acquired by the users as part of increasing personal skills in management. Another part will be acquired from an external supply (for example, an extension service) with its associated cost or price. This raises the question of the ability of users to acquire or pay for their information needs: if not in money terms, there is at least the time and attention to be paid by farmers for involvement in informational events which may otherwise appear to be completely

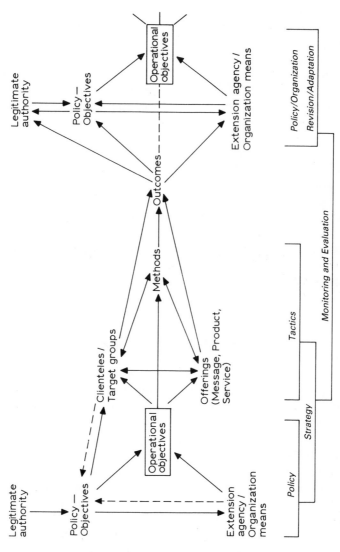

Fig. 1. A dynamic view of objectives as the central element in the extension process.

provided by public funds. Effective access to, and effective demand for extension services are difficult to measure.

If agricultural information is to be seen as a factor of production, it stands both as an independent factor and in interaction with land, labour, capital and managerial ability. Land may be improved in quality (for example, fertility) by the knowledge and skill brought to bear on its use: it may also, of course, deteriorate. Both quantity and quality of labour are influenced by formal education and training at the recruitment stage, and by non-formal education (including extension) thereafter. The productivity of capital investment is highly dependent on its management, and especially the abilities to allocate and alter the total resource mix available for production from the basic (renewable) natural resources. Farmers' management abilities can be enhanced not only by experience but also by better knowledge and relevant information. In common with other factors of production (except usually capital), knowledge may appreciate for a period of time as it gains incremental value from use and experience, before it begins to depreciate as a first step towards obsolescence.

The extension services are thus to be seen as part of the knowledge system for agricultural and rural development, and as partly responsible for the establishment and management of that system. Moreover, this will have to be seen as an extremely complex system, or set of systems, of an overall open nature. The management of the extension service component of this fluid knowledge system has increasingly been improved by the application of some general principles that have emerged from management in the private, commercial sector, such as product promotion and marketing, and by specific systems such as T & V. If this is to be taken further in a rational way, some clear criteria are urgently needed on the basis of which the opportunities for investment in improving the contribution of extension to rural development can be clearly identified and subsequently evaluated. These criteria are many, and will vary according to specific locations, countries, and culture. They should certainly include: the size and content of the existing fund of information for immediate dissemination; the location; the nature of improvable farming systems; the character of the total system of channels of communication available for use with rural people; the adequacy of the links between research services and rural people; and the extent to which the strategies and goals of extension work are adjusted to the constantly changing new contexts in which it is required to deliver a good or at least adequate rate of return on the investment involved. How this investment should be distributed between potential clients (all farmers or some, all rural people or only those who might or should benefit), and between the public and private sectors (all industry, agro-industry, national and international) is another key question.

CONCLUSIONS — AND SECOND INTRODUCTION

What has emerged from this review of the dimensions of rural extension can now be summarized. Disseminating research results is a long established and important role. Relevant, reliable and useful knowledge and information concerning innovations and improvements is of great value to agricultural production. Beyond this, however, is the information role in agricultural services such as marketing, credit and farmer organizations, and in rural services, facilities and organizations, and in public involvement in many aspects of the rural environment. Of increasing concern, as an issue, is the extent to which what have sometimes been called 'clients' or 'clienteles' can participate *actively* in solving their *own* problems with the help of extension workers functioning as advisers rather than persuaders. This leads to new views of the scope and functions of extension workers and of their roles not simply as information purveyors but as guides and counsellors to local sources of valid information. In turn, this also implies new uses for media and information technology which are rapidly changing and developing. There are obvious implications for the training of extension workers, and for the numbers of staff required and how they can be most effectively deployed.

There are also clear implications in relation to the cost of extension as a public service, with a consequent need to accept (and if possible overcome) the difficulty of assessing and making a clear specification of the returns and benefits which accrue. The issues are primarily those of management concerning the broad policy of extension, and the extension personnel and their work, using in this process improved (probably not yet available) methods of monitoring and evaluation. Of most importance is the quality of the help given to farmers and their families, and to rural people, in achieving their aspirations.

Discussion

DONALD C. PICKERING*

Assistant Director, Agriculture and Rural Development, The World Bank, Washington, DC, USA

The paper by Rolls, Jones and Garforth sets the scene well for considering the topic 'Investing in Rural Extension'. It provides insights *in various countries* into the origins of extension, the development of extension services, the dimensions of extension, with some challenging thoughts on new dimensions, and ~~very appropriately,~~ a section on investing in extension. Generally, I found myself in reasonable agreement with most of the views put forward.

This said, however, there are a number of points touched upon in the paper that deserve more powerful emphasis. The first, and without question the most important, is that in developing countries generally, agriculture is the pre-eminent economic activity. With the exception of the oil exporting countries, which are mining a declining rather than utilizing a self-sustaining or a potentially expanding resource, most developing countries are heavily dependent on their agriculture for the progress of their economies and hence the well-being of their people. Contrary to the conventional wisdom of the mid-part of the twentieth century, the starting engine for economic growth is now recognized to be agriculture. Unless a country's agriculture is healthy, dynamic and capable of meeting domestic needs directly or indirectly, via the foreign exchange earnings needed to purchase the agriculturally derived and other products called for by domestic demand, the prospects for economic development are, in most cases, far less than fully satisfactory. Hence, the importance of the agriculture sector, and hence, by simple

*The views expressed herein are the author's alone and in no way should be construed as official views of the World Bank.

19

inference, the importance of increasing the productivity of the individual producers in the sector.

Increasing the productivity of the individual producers can be achieved by any one of a number of interventions in terms of national economic development strategy, but experience has shown that combinations of such interventions lead to the most cost-effective approaches. Before anyone in a Ministry of Agriculture, or in Research and Development anywhere begins to talk about the organization of agricultural, or indeed rural, extension they need to have a very clear understanding of overall economic development goals and the strategy that government proposes will be employed in meeting these goals. Thereafter, it should be possible for policy makers to determine and decree the directions to be taken by national research systems as to individual crop and livestock focus, on the basis of political considerations and also, one hopes, a large weighting on estimated comparative advantage in production.

I would argue that very few national economic development policy makers are sufficiently aware of the constraints imposed upon agricultural research workers by the natural resource endowment and perhaps, more importantly, the human resource endowment in their countries. Although the authors of the paper (Rolls, Jones and Garforth) have touched upon these issues, in my view they have not paid sufficient attention, or given enough emphasis to this broader canvas of economic development which clearly is facilitated by effective agricultural research and extension. In short, my call is to put the effective organization and management of agricultural extension into the context of national economic development where it belongs, together with agricultural research.

This said, I admit that I speak as a person who views agricultural extension from some little distance, which may be no bad thing. On the other hand, having run a national agricultural extension service for several years in the 1960s and subsequently taken a broader view though still in the agricultural sector, I think I can claim to have an understanding of the problems that we have to debate. And so to more specific aspects.

With regard to the development of extension services, I commend the focus on management, alluded to in the paper with particular reference to Training and Visit extension. This allusion is entirely appropriate. Training and Visit extension is clearly a dynamic management system. It is not, as some critics would have it, 'Holy Writ' transcribed from a tablet brought down from some mountain, and not to be questioned or modified to meet local situations. It is rather a set of commonsense principles that, admittedly, requires intelligent managers to review and adapt to their particular social, political, and economic conditions. These principles can be intelligently and successfully applied in economic terms in agriculture. However, if they are not, then criticisms of high cost and lack of effectiveness may be justified. The evaluation work of Feder, Lau and Slade (1985) in Haryana State in India demonstrates clearly the effectiveness of the Training and Visit system compared with the more traditional mode practised in India in terms of technology transfer and economic rate of return.

Like any activity that has to do with development, and certainly rural extension has much to do with development, the system practised must evolve to meet changing circumstances if it is to continue to be useful. As additional technology is developed, and as farmers become more knowledgeable in the application of new technology, the role of the extension worker as the link between the farmers and the researcher becomes increasingly complex, whether contacts with either are direct, via personal visits and demonstrations, or indirect via media messages and the like. Recognition of the need for evolution of extension systems by their managers is essential, and research into this evolutionary process strikes me as being a somewhat neglected field. Although the World Bank is not an institution that is particularly noted for research outside the field of development economics, we are concerned with the high recurrent costs, in terms of human and other resources, of many extension services that we have helped and are helping to finance. This is particularly apparent in some of our poorer client countries, and for this reason we are proposing to investigate the impact on the efficiency of extension systems of reducing the recurrent cost burden, by such means as lowering the agent density and increased use of the media.

As a footnote to the paper's mention of the increasing application of both the concept and practice of extension outside the confines of agriculture, I would mention a World Bank publication on adapting the Training and Visit system for family planning, health and nutrition programmes (Heaver, 1984). This emphasizes that a 'model' system proposed for population health and nutrition programmes is not a blueprint but rather a starting point for further adaptations appropriate to local conditions. It looks at three such systems on the ground in India, the Philippines and Indonesia, and suggests alternative approaches to making such programmes more cost effective, outlining some of their theoretical advantages and disadvantages. Also being tried in Kenya is the application of Training and Visit principles to the provision of credit services to small farmers, with claimed satisfactory results.

Concerning the dimensions of extension, the paper rightly draws attention to the increasing recognition that research services are not omniscient and that their recommendations are not invariably appropriate for all, or sometimes any, of the farmers to whom they are directed.

I have already emphasized the crucial importance of agriculture in the economic development scenario of most countries, and within this point is the importance of ensuring that research services perform work that is relevant to this need and relevant to the needs of the farmers they serve. This view is receiving increasing recognition by policy makers and research managers in many — but far from all — developing countries. The rise to prominence of Farming Systems Research in the programmes of both national and international research organizations is one manifestation of this fact. The paper by Simmonds (1985), commissioned by the World Bank, provides useful insights into this subject, though some would wish for a significantly more even-handed treatment of the sociological issues involved. The diagnostic survey approach to farming systems research propounded and practised by Collinson of CIMMYT emphasizes the importance of linkages between the researchers and the farmer. Such linkages, and the role of the extension worker in creating and strengthening them, received close attention from both research and extension managers, principally from Asia, at a workshop jointly sponsored by UNDP, FAO and the World Bank in Indonesia in 1984 (Cernea et al., 1985). This theme was also followed, though with a greater emphasis on cost effectiveness, at a West African workshop at Yamoussoukro in Ivory Coast in early 1985 (proceedings forthcoming). This was especially interesting in its consideration of the evolution of extension systems and, in particular, the blending of the commodity-driven system pioneered by the French Company for the Development of Cotton Textiles (CFDT) with key elements of the Training and Visit system and/or the more participatory groupement villageoise approach in projects financed by the World Bank in Western Africa.

In their paper, Rolls, Jones and Garforth caution us to remember that research and extension services are only a part of an approach to change. They note that the policies and institutions responsible for their formulation and implementation inevitably operate at a distance, geographically, from the rural communities they serve. I would add to this the observation that in many cases, perhaps far too many cases, the institutions and their staff operate at even greater distances attitudinally from the rural communities they serve.

One cannot but agree with the contention that the need is to focus attention on the social and economic affairs of rural society with increasing invention, adaptation and management of technology by the members of that society, rather than by outsiders. As a representative of an organization that is much more of an outsider to rural society in any country than that country's institutions, I take the point very clearly. Unfortunately, one cannot say the same of all one's colleagues. However, there is an increasing realization of the validity of the concept of maximizing beneficiary participation in all development interventions if they are to be sustained. It is a logical next step, having involved beneficiaries in the planning and management of interventions designed to help them, to move from the strategies of change, as postulated in the early T & V model, to strategies of choice, although as the authors of the discussion paper admit, there is a blurred distinction between the two.

I wonder if the authors make rather too much of the heterogeneity of rural society as a limiting factor to the impact of the 'research/ technology/dissemination approach to the promotion of rural development'? Whether they do so or not is immaterial, however, provided one establishes the nature and extent of such heterogeneity as may exist as a prelude to the detailed design, planning and implementation of the research/technology/dissemination approach. Indeed, this investigation of rural society should be the first part of the 'research' link in the approach. Thereafter, one can establish target groups and with them their particular needs. Unfortunately, this happens all too rarely.

I would argue that, having taken such a step as outlined above, much of what the paper by Rolls, Jones and Garforth recommends should follow naturally. Having established levels of communication with the target groups, evaluated their needs and capabilities, designed technologies that are appropriate for them and sensitive to national or regional development goals, and provided the means for the efficient dissemination of the technology and the feedback from farmers for such technology modification as may be needed, one will rapidly reach the stage where the extension worker functions as the adviser. It is not a far cry, thereafter, provided the advice is seen to be good, for the farmer, given the choice of paying for the advice or not receiving it, to demonstrate the extension service's cost effectiveness by his willingness to pay for it, as for any other service which is perceived to have a monetary value. We are still a long way from this happy state of affairs, but I am convinced it is an attainable goal.

CONSTANTINOS L. PHOCAS
Director, Department of Agriculture, Cyprus

I would like to start my discussion of the paper by Rolls, Jones and Garforth by clarifying the content of the concept of agricultural or rural extension. Some people tend to equate it with the term *technology transfer*. This is incorrect, because technology transfer includes the additional functions of input supply and agri-services. In addition, extension includes training farmers in agricultural and home management and decision-making skills, as new technology inevitably places more demand on these abilities. Also, agricultural extension helps rural people to develop leadership and organizational skills, so that they can better organize, operate and/or participate in cooperatives, credit societies and other support organizations, as well as participate more fully in the development of their local communities. Therefore, while agricultural or rural extension is an essential and major part of technology transfer, the terms are not synonymous.

The paper under discussion will set a comprehensive background not only for the present purposes but also for extension work in the world beyond the twentieth century. Having said that, I submit that 'rural' or 'agricultural extension' is in most developing countries, not to mention many developed ones, in a state of flux today. Although governments all over the world recognize the need for an effective extension set-up and work, it is mostly depressing to realize that in a great number of countries the role and practice of extension has to start in an entirely fresh way, if it is to meet its goals in social and economic development. This is no exaggeration if we consider how many of the following ten fundamental prerequisites for an effective extension system exist in the majority of the countries in the world today.

(i) Effective linkages with research organizations.
(ii) A statutory basis and mission.

  (iii) Stable financial support.
  (iv) On-going and effective in-service training.
  (v) Adequate field offices, transportation, and communications systems.
  (vi) An orientation to understand and to serve the farm family.
  (vii) Freedom from regulatory or input supply responsibilities.
  (viii) Continuing farmer input to guide programme priorities.
  (ix) A competitive salary system with incentives for professional development and advancement.
  (x) An effective communications/information staff.

The picture of the world map of extension systems on the basis of the above criteria is very gloomy indeed.

This depressing situation is the result of a mix of factors, most prominent among which are poor information on the part of many decision makers in governments, inadequate management, social and economic policies pursued by governments, and limited human and financial resources. The social and economic policies pursued by governments constitute the focal point of the above complex, and consequently this is a decisive factor as regards the shape, the strength and the orientation of the respective extension systems.

Does extension always serve the most needy farmers, or is it the case that, because of the existing institutional infrastructures and economic structures, or because of governments' economic policies to achieve maximum and fast-yielding returns from their investments in the agricultural sector, extension systems are geared towards serving the bigger and more affluent farmers rather than the small ones? Let us not forget that large numbers of small farmers are difficult to reach because of limited human and material resources on the part of the governments (limited numbers of extension staff, poor logistic support, etc.) and poor communications; and small farmers are usually low in receptiveness because of their low literacy rate, inaccessibility to credit and other inputs, and generally their low adoption rates.

Therefore, corrective action should be taken to ensure that extension is primarily oriented to the small farmers if social considerations are attributed equal priority to the economic ones, and if more balanced distributions of wealth, equity and equality of social rank within a nation's rural community are to be achieved. The 1979 World Conference on Agricultural Reform and Rural Development (WCARRD) contributed substantially in boosting the above policies.

In many developing countries a prevailing fallacy tends to undermine the achievements of agricultural research, implying that the extension services are left with nothing to extend to the farming communities because of poor performance on the part of agricultural research. Although there may be some exceptions, the problem is rather one of teaching and transferring available knowledge to the majority of farm producers on the one hand, and on the other communicating to the research institutions the problems to be addressed by research scientists. In this respect, the 'Farming System Approach' is acquiring ever-increasing importance, and its ultimate objective is to maximize farm productivity and income. Effective research–extension linkages, which are mostly neglected, should therefore be given appropriate attention and be attributed high priority by addressing such topics as:

(i) the identification of farmers' production problems;
(ii) the generation of agricultural improved technologies to solve agricultural production problems;
(iii) the identification and formulation of extension technical packages; and
(iv) policy and institutional issues related to research and extension linkages.

It is pertinent in the above context to quote the results of a recent study in the USA, where 1400 agricultural scientists were asked to rank the criteria they use in selecting problems for research. 'Enjoy doing this kind of work' was given the highest rank; 'importance to society' came in second; while 'funding' was ranked ninth; and 'feedback from extension personnel' was ranked twentieth, out of twenty-one possible criteria.

'Starting from the farmer's level' is a motto in extension work. Therefore, reference made in the paper under discussion to the need for research work to adapt its procedures and its understanding of farming systems to indigenous production systems is highly relevant. But the authors of the paper have gone a step further in suggesting an innovative approach, whereby attention should be focused on rural society rather than outside agencies as the initiator of the process. Nevertheless, the authors, conscious of the implications of such a process, are quick to emphasize possible major obstacles to its adoption. Time and again it has been stated that extension's major function is a 'catalytic' and accelerating one. Therefore, given extension's key role in the agricultural, economic and social development of every country, the need and importance to refer to a 'farmer–extension–

research continuum', with varying degrees of managerial autonomy between research and extension, acquires ever increasing importance. Furthermore, the agricultural development effort of any given country should be seen as a 'system' made up of sub-systems which, in addition to the agricultural research–extension sub-system, include credit, input supply, communications and marketing sub-systems to mention the more fundamental ones.

In the above context adaptive research, which would address farmers' production and management problems with the aim of maximizing farm income, acquires high priority within a country's sectoral agricultural policy.

Reference in the paper on the emphasis in supporting the development of self-sustaining groups as a countervailing force in rural society, through which powerful individuals, groups and institutions can be confronted, merits special attention. Assessment of the degree of effectiveness of such a function in extension work would contribute materially in furthering its scope and programmes. A valid question to ask, however, would be whether such a function would be realistic and effective especially in those cases where the extension organization is part and parcel of the governing socio-economic establishment of the countries concerned.

Criticisms of extension effectiveness should rather aim at improving the management and institutional framework, and the component sub-systems of inadequate marketing, credit, input supply and communications infrastructure, instead of placing the blame on extension *per se* of being too narrow and unambitious or, at other times, of trying to do too much with inadequate resources and staff.

An interesting approach to extension services is made in the paper as being both part of the knowledge system for agricultural and rural development, and partly responsible for the establishment and management of that system. This is a comprehensive statement which underlines the very special nature of extension work as a concept which embraces both the *object of management* and the *art of management* in one indivisible highly dynamic and self-sustained process.

The information role of extension in agricultural services such as marketing, credit and farmer organizations does not need to be over-emphasized. Nevertheless, extension's role in the *organizational aspects* of such fields should be given equal priority if effective action and sustained results are to be expected. Therefore, the teaching function of extension should, wherever possible, be supplemented with an effective leadership role in sustained organizational and operational activities.

The role of the media in the diffusion process in extension work has been thoroughly covered in the paper. One wonders, however, with the ever-accelerating technological developments in this field, whether there will be a limit to the use of the media in extension work in the near future?

Finally, a topic not given appropriate prominence in the paper is extension work with farm women, mainly as farm producers and as farm and rural home managers. In a world where more than half the active agricultural population consists of women, there is an urgent need to plan and initiate intensive extension work specifically geared to addressing the multicomplex problems faced by farm women. Rural development based on growth with equity will require the full integration of women, including equitable access to land, water and other natural resources, inputs and services, and equal opportunity to develop and employ their skills. There is also an urgent need to expand knowledge and statistical data on all aspects of women's roles in rural activities, and to disseminate this information in order to promote greater awareness of women's role in society.

A major turning point in FAO's approach towards and programme of work on women was the 1979 World Conference in Rome on Agrarian Reform and Rural Development. It recognized women's role not only as farmers and members of the community at large, but also the importance of consulting and involving them in the decision-making process with the aim of reflecting their views more effectively in rural development strategy.

In the light of all the above, it is imperative that extension services (agriculture and home economics) including access to inputs, market, credit, technology be reoriented, with men as well as women being trained *how* to reach women in culturally accepted ways. Therefore, we need:

— to establish special recruitment and training schemes to increase the numbers of women in the training and extension programmes, including professional fields from which women have been traditionally excluded;
— to broaden the range of agricultural training and extension programmes to support women's roles in activities of agricultural production, processing, preservation and marketing;
— to promote income-generating activities for women;

— to strengthen, through extension programmes, the training
opportunities for rural women, including leadership training,
instruction in agricultural as well as non-farm activities, health
care, upbringing of children, family planning and nutrition;

— to ensure that technological innovations should relieve the
drudgery and heavy burden of women's household and farming
activity, but should not reduce their income earning opportunities;
and

— to promote home economics functions within extension services.

What has been referred to above is in no way exhaustive of the many
discussion issues which are raised in the paper. I have tried to present
my comments, mostly in a concentrated form, on key issues in extension
work. They are those perceived by an extension worker who has gone
through all stages of extension and rural development management, in
areas and countries covering a wide range of the agricultural, social and
economic development spectrum, as well as geographic location, over
more than a quarter of a century of professional life, and in perspective
looking into the twilight of the twentieth century and the dawn of the
twenty-first century.

PAUL INGRAM
*Senior Agricultural Officer, ADAS, Ministry of Agriculture, Fisheries and Food,
London, UK*

INTRODUCTION

It is to be expected that the dimensions of rural extension differ with
location and circumstance, particularly according to the understanding
that farmers have of the value of advice. In the United Kingdom,
changes in circumstances have occurred which are persuading us to
take a new view of both our agricultural industry and the extension work
of the Agricultural Development and Advisory Service (ADAS) in
England and Wales and of the corresponding advisory services in
Scotland and Northern Ireland.

Taking the ADAS as an example of these services it has developed
over a period of nearly forty years on classic extension principles. There

are now in the ADAS some 4500 staff, about 2000 of whom devote some part of their time to advisory work. The purpose of this work has been and is technology transfer using two patterns of contact with the industry. One of these has been to provide a consultancy service, making advice available on request to all farmers. The second, which has increased in importance with the recent development of the planned advisory project, has been pro-active with the subjects for advice being chosen by the adviser. There has been only a limited amount of participative activity involving a partnership between farmers and advisers in deciding what technologies are important and how they should be developed and applied.

Just as farmers have not participated to any large extent in the formulation of advisory or for that matter research and development programmes, neither have they been required to make any direct contribution to the cost of their provision. The ADAS has so far provided free advice on request to all farmers, the objective of which has been the personal benefit of the recipient. The result of this advice has been both increases in the efficiency of production and increases in the total level of agricultural output.

CHANGING PERCEPTIONS IN THE UNITED KINGDOM

In Western Europe as a whole, but articulated particularly in the United Kingdom, perceptions of agricultural policy and agricultural extension are being changed by a circumstance which is new to us. This is the emergence of surpluses of virtually all the major agricultural commodities. This situation produces a dilemma, as yet unresolved, in relation to rural policy. Though in most of these countries there is a substantial rural population for whom governments would welcome an improvement in income, any simple product pricing policy which achieves this end will also stimulate the production of further surpluses.

For the extension services there is a parallel dilemma, though often not one so clearly exposed. This concerns the historical emphasis of these services on advice which improves farm economic performance and usually also levels of product output. Now, it seems possible that to continue with this kind of advice might no longer be such a good national investment and that the direction of the extension services' advice should perhaps change. Greater emphasis might in future need to be given to systems of production limited by specified output requirements which are not defined by unit product price alone, and sources of non-agricultural income for farmers might assume much greater importance. Such changes would imply for the ADAS new dimensions in agricultural extension which are largely outside our present experience except for some inputs into rural socio-economics.

Quite apart from these macro-economic pressures on agriculture as a whole and the effects that these might have on our extension policy, the United Kingdom government has recently been considering the extension services themselves and their relation to the industry. The government must see, as would any other observer, a technically efficient, well modernized, well-informed industry. This is, however, an industry which is highly supported both by European Community and by national measures, the level of this support being close to the net income of the sector as a whole. In addition, the government provides some £220 million a year to pay for research, extension and certain statutory services.

As part of this consideration of the extension services, in 1984 the Director General of the ADAS produced a report which suggested that it would not be unreasonable for farmers to pay some part of the cost of the provision of advice (Bell, 1984). Given the circumstances of the industry, coupled with a general desire to reduce government expenditure, it is not surprising that this suggestion has been adopted. It has now been translated into financial targets for the ADAS, and for the extension services in Scotland and Northern Ireland, for the year 1987/88 which imply either financial contributions from farmers or a reduction in the size of the advisory activity.

The questions which face the ADAS as a result of these proposals are among those identified in the paper by Rolls, Jones and Garforth. They include:

— should the emphasis of style be less persuasive or coercive and more collaborative and participatory?
— is the classic transmission of technology from research to the farmer the most important future role?
— should farmers have more influence on the objectives of research and advice and, if so, then how?
— what communication systems are now most appropriate for extension work in the United Kingdom?
— what balance is desirable between government investment in extension and financial contributions from the industry?

Fortunately, in attempting to answer these questions, it is possible to look at extension services elsewhere which, because they have developed different systems of operation, can now provide models for possible changes here. This applies particularly to the extension services of our nearest neighbours in France, the Netherlands, and Scandinavia.

MODELS FROM EUROPE

A survey of the European scene discovers some approaches to extension work which are rather different from those used in the United Kingdom. In comparison with the United Kingdom we see in France, the Netherlands and Scandinavia high levels of farmer participation in the organization, running and financing of the advisory provision. In some cases there are also interesting developments in the use of computer-based information systems.

France

Since 1966 agricultural advisory work in France has been taken over entirely by organizations within the industry. The responsibility for carrying out advisory programmes in Departments and Regions lies with the Chambers of Agriculture and most of the advisory work is undertaken by independent agricultural groups employing their own advisers. These groups include: agricultural advisory groups of 15 to 150 farmers; farm management and rural economy groups of 100 to 500 farmers; cooperative organizations which at times provide advisory services for their members; and groups of young farmers and wage earners in farming.

The advisory work is financed by subsidies from either central or local government to meet about 24 per cent of costs, from levies organized by government on products such as cereals and sugar beet which meet a further 25 per cent of costs, from contributions by Chambers of Agriculture, and from direct charges to producers for specific services. The overall provision of advice is supervised by the Association Nationale pour le Development Agricole (ANDA). This body has on its Board of Directors representation from the Chambers of Agriculture, Farmers' Unions, Young Farmers' Unions, Agricultural Cooperatives, Mutual Credit Funds, and other associations for specialist branches of farming.

The Netherlands

In the Netherlands, government policy has been moving towards a greater degree of self-help by the industry in the provision of advice. Since 1983, a structural reorganization of the state advisory service has coincided with an increasing emphasis on the use of extension clubs, to each of which an extension worker is associated on a regular basis. Government strategy, it seems, is to be less paternal with a move away from the routine visiting of farms and a lower level of provision of routine prescriptive advice. The Farmers' Unions cooperate closely with the State Advisory Service in the organization of the extension clubs, and they have also formed societies for the Technical and Scientific Improvement of Farming which are also encouraged and supported by central government. It is estimated that about 80 per cent of all producers belong to such societies which are also used as a medium for advice. The societies are run by local Boards of Directors and funded by subscription.

In addition to this already close relationship between farmers and the state service, the Farmers' Unions also employ some 65 specialist advisers in pigs, arable crops, dairying and horticulture, and 220 social and economic advisers all of whom are financed half by the unions and half by government.

Denmark

The Danish advisory services are entirely managed by the farmers organizations, in this case the Federation of Danish Farmers' Unions and the Danish Smallholders Union in cooperation with such bodies as the Artificial Insemination Societies, the Recording Societies, the various National Breed Societies and the Bacon Factories. The State provides a subsidy of 70 per cent of the salary and travel expenses of an agreed number of advisers and their support staff which represents about 25 per cent of the total costs of the advisory services, the remainder being met by the agricultural organizations and charges to individual farmers.

At field level, advisory work is carried out through about 150 local unions from which representatives are elected to serve on local and national management boards. It is striking that the Danish farmer would on average, say for a unit of 40 to 50 hectares with either 100 sows or 50 dairy cows, pay annually fees of about £1000 for the provision of advice. In addition to managing advisory services at field level, the unions also finance the Danish Agricultural Advisory Centre (which guides development work and provides specialist advice and literature), the Agricultural In-Service Training Centre and, in cooperation with the AI and milk recording associations, a central electronic data processing (EDP) centre.

Sweden

The Swedish State Advisory Service also works closely with the industry through management boards of farmers in each of the 24 advisory areas. Features of particular interest here are recent moves towards charging individual farmers for the advisory services they receive and the development of a videotext information system by the industry.

Farmers in Sweden are highly organized into cooperatives which themselves employ advisers alongside whom, and in association with whom, the state advisers operate. A recent change in the nature of the state system is that as from April, 1984, the Service had to accommodate a reduction in state funding of 5 per cent over a three-year period and, in order to maintain the level of advisory provision, raise a similar sum by the introduction of charges for advice and services. This they have been able to do, beginning in July, 1984, meeting the equivalent of the whole 5 per cent reduction in that financial year. These targets have been achieved with a high degree of autonomy at local Board level as to what charges should be imposed and at what rates.

A significant development in Sweden is the central electronic data processing facility provided by a combine of the Federation of Farmers, the meat marketing organizations, the supply and marketing cooperatives, the cooperative banks, and the Swedish Association for Livestock Breeding and Production (which is itself a combination of 20 livestock cooperatives, five semen producing cooperatives and the breeding societies for cattle and pigs). The facility has a computing centre supporting some 2500 terminals of which about 300 are now on farms, the remainder being distributed among the participating bodies. Accounting, recording and management services are offered with the ability to carry out transactions such as banking and amendment of records using a recently introduced videotext service. It is hoped to have 10 000 farmers on the videotext system by the early 1990s.

Finland

The Finnish extension system shows a high degree of farmer involvement and financing, and widespread use of computer-based information methods. Advisory services are provided by the Association of Finnish Speaking Farmers for the majority of producers. It operates through 15 provincial agricultural centres in which there are 300 agricultural societies made up of 3500 agricultural clubs with 250 000 farm members. This structure provides the organization of the service at local, regional and national level. There is no government involvement in management even though the government pays 45 per cent of the total cost, the remainder coming from membership subscriptions and fees charged for services and advice.

A notable development in Finland is the use of 600 portable computer terminals by advisers. These are supported by a central processing facility and a further 200 terminals will come into use in the autumn of 1985. In the last full year of operation, between 35 000 and 40 000 farmers made 75 000 uses of services provided by this means. The services used were as follows:

12 000 — fertilizer planning,
20 000 — concentrate feed planning,
23 000 — basic feed planning,
9 000 — economic milk production,
4 000 — fodder production planning,
3 000 — taxation and accounting,
4 000 — animal breeding,

in addition to the use of the facility by the AI organization, for milk recording, by dairy factories and by slaughterers.

CONCLUSIONS

This brief comment on the challenge facing the advisory services in the United Kingdom, and the examples of the way in which similar challenges have been met elsewhere, has lessons for the ADAS, and lessons which can perhaps be applied more generally.

It is difficult not to be impressed by the successful way in so many countries in which the management of the advisory services has, in the words of the introductory paper, seen, 'the authority for intervention in the social and economic affairs of rural society as residing within that society itself rather than with policy makers and administrators'. At the same time it must be said that this authority has some limits imposed by government policy whether on product prices, farm structure, the freedom to change the nature of the countryside and so on. It also must imply, and is seen as doing so in those countries cited as examples, the assumption by the industry of some financial responsibility to go with the authority to determine its objectives and organization.

Anyone examining the advisory services in the countries which have been given as models will see that they are by no means as simple as this brief survey suggests, nor perhaps as effective as some idealized participative system might be if designed with no existing constraints. But, as the introductory paper makes a point of saying, no services, in Western Europe at least, are able to start in an entirely fresh way and have to be developed from some historical base. What such an examination does reveal is the large number of ways in which farmers can participate in and fund their own extension services; and it supports the view that whatever the nature of these services, whether concerned only with technology transfer from research or having some other role, such participation does strengthen relationships between advisers and farmers.

The use of new communication systems based on computers is less developed in these countries than their established systems of participative advice based on face-to-face contact. They provide an example of three ways in which information technology (IT) can be used. In Denmark there is a highly developed central processing unit and all advisory centres are equipped with terminals. In Sweden a similar central unit with an impressive interactive videotext facility is operated by the cooperative organizations with the intention of establishing terminals on farms. In Finland a similar service has been developed by the advisory service which has equipped advisers with terminals. This third approach, that of making IT available to farmers, together with the trained operational and interpretative skills of an adviser, seems to have much to recommend it and has certainly been effective in getting IT into use on farms.

As far as the questions facing the ADAS are concerned it is possible to give well-intentioned responses if not yet to say what mechanisms might be possible to put these intentions into practice. The messages for the ADAS in England and Wales are probably equally valid for extension services elsewhere, at least as longer-term goals. They are to accept the value of collaborative and participative approaches to extension work; to accept a role which is wider than the mere transmission of technology; to accept the value of farmer influence on research and advisory goals; to recognize the value of new IT systems if introduced in an effective way; and to expect the participation of farmers in the financing of the services which they use.

2

Extension and Technical Change in Agriculture

A. H. BUNTING

Faculty of Agriculture and Food, University of Reading, UK

I propose to consider agricultural extension from two points of view —
its relation to other parts of the agricultural knowledge system, and its
relation to other factors which induce or oppose progressive change in
agriculture and in the rural space. I shall then go on to consider the role
of extension in general. Along the way I shall attempt to persuade you
that the success or failure of extension may be determined by weaknesses
elsewhere, and that it is far from easy to isolate the effects of extension,
so that though we may define strategies and goals, the return to
investment in extension may still remain inherently difficult to
measure.

EXTENSION AND KNOWLEDGE SYSTEMS

Extension is customarily seen as a means of transmitting knowledge to
producers. As the theme of this volume suggests, knowledge is a
resource for agricultural production alongside capital, land and labour,
all of which, in formal terms, it helps producers to extend.

Knowledge and Knowledge Systems

However, knowledge is a complex resource. It includes information,
concepts, techniques and skills, which are maintained, increased,
disseminated, tested and used in knowledge systems. However disorganized and unsystematic it may seem, the agricultural knowledge
system, whether local, national or international, has five essential
components.

37

The first is the *existing stock of knowledge*. Perhaps the most important part of that stock is the knowledge which is stored in the minds and memories of men and women, particularly those who depend for livelihood and survival on biological production; it is a product of long individual and collective experience of environment and living organisms. Other essential parts are held in libraries, books, journals, maps, survey reports, filing cabinets and archives, and retrieved through increasingly effective abstracting and indexing systems.

The second is the *means of increasing knowledge*, by experience, surveys and above all by experimental research, which is intended to obtain new information, test critical hypotheses, put knowledge into acceptable order by means of concepts, and improve techniques and skills. This component includes research on farming systems themselves, and on the life systems of which they are a part; it is intended to describe them, analyse their rationale, and assess their resources and constraints in order to indicate possible options for more productive change.

The third component is the means of *testing and developing knowledge,* the development part of research and development, so as to fit it for practical use in specific circumstances. This component includes the so-called on-farm research, in which research station results are tested under allegedly farm conditions in partnership with producers. Some would say that it is all too often an excuse for drawing promotional conclusions from bad experiments, but that judgment is too harsh: after all, both British and Indian fertilizer and breeding policies were developed, and continue to be tested, by large numbers of thoroughly respectable experiments conducted in collaboration with farmers and in their fields and crops.

The fourth is the *practical application* of new and improved methods and processes to increase output, lessen costs of products, and adjust the production system.

Finally there is the *dissemination* component, through which knowledge passes between participants in all parts of the system to help them to work more effectively. This sector includes education and training of many kinds as well as the various activities we group together under the name of extension. In addition to their many other tasks, universities have a key role in this component: the strength of the universities in a nation determines the strength of all the other parts of the agricultural knowledge system, since they are led and staffed, at least in part, by university graduates. If the universities are weak, no other part of the system can be strong.

Weakness in any of these five components is likely to be reflected in unsatisfactory returns to extension. The most common weakness (outside the extension organization itself) is in the knowledge-increasing component, and particularly in research, which in the protected conditions of the research station may produce results which are not relevant to the needs, resources and difficulties of real producers, or to the requirements of national governments. In not a few nations the universities are weak also.

The Uses and Content of Extension

Agricultural extension appears to have been seen in its early development in the formerly virgin lands of the Middle West of the United States as a means of extending the work of the new agricultural colleges, and the results of their research, outside the academic walls, from the scientists to the producers. However, as the years have passed and knowledge and understanding have grown, it has evidently become equally necessary to transmit knowledge in the opposite direction — from producers and other rural people to scientists, research workers, managers, and those who form policy. All of these need to be sufficiently informed about the objectives, resources, constraints and other difficulties of producers. Perhaps this is why, in so many developing countries, extension is a politically sensitive field: in one nation which I know well it is attached, not to the Ministry of Agriculture, but to the Office of the Prime Minister.

Customarily, I believe many people have seen extension workers as passive intermediaries who receive output from the rest of the knowledge system and communicate it more or less efficiently to those who are thought to need it. But it seems increasingly evident that the effective extension worker must be able to codify, to synthesize and to interpret knowledge, and to process it into forms and options which are useful to, and acceptable by, all his many actual and potential clients. He is an active participant, not a neutral, one-way conduit.

Perhaps, therefore, extension is not the right word to describe the task. We do not use the term for the official or private sector services in Britain. It may well convey an inappropriate idea, but I suppose it is too late in the day to change it now. After all, the word has been hallowed and respected elsewhere for at least 100 years. But let us not forget that it may perpetuate a counter-productive image of the nature of the work among the many non-professionals who have to guide national and international development policies and investments.

The Transfer of Technology

For me the phrase *transfer of technology* resembles too closely a sales promotion gimmick. The phrase has been imported illegitimately from the discussions in international forums, such as the United Nations Conference on Trade and Development, about the transfer of patented or secret industrial technology from developed to developing countries.

In relation to agriculture, the words carry the conception of the scientist, 'Big Brother', who knows or can find out all the answers, handing down guidance and direction, through extension, to the grateful and expectant producers, an ignorant and reactionary lot who have to be induced to use the goodies generated by 'Big Brother' in order to make progress. This carries a top-down, centralized, authoritarian air, an assumption that the scientist commands a prefabricated technology which the producer must somehow be persuaded to adopt, whether he likes it or not, because the result is expected to serve the policies of government, and may even be good for the producer as well.

The phrase may also conceal an old colonialist implication that because the developed countries know how to do the job, all that is needed is adaptive research to make western knowledge suitable for totally different environments.

Both these concepts have long since ceased to attract me, for reasons some of which will appear below.

I can sum up so far by saying that extension is but one part of a complex system for the management of knowledge, and that its strength and effects are likely to be strongly affected by other parts of the knowledge system. Further, it is not the task of extension to give orders to rural people: its job is to help them in their efforts to achieve a better life — efforts which are the main engine of development in agriculture and the rural space.

EXTENSION AND CHANGE IN AGRICULTURE

I propose next to consider some experiences of technical change in agriculture in order to indicate the contributions which extension has made to them. I start with some queries about the measurement of change, without which we cannot assess the return to investment; and then I shall consider the many factors other than knowledge which affect the change process.

Any attempt to measure the effect of a single element on change in a complex and changing system is likely to encounter the logically insoluble difficulty that, even when the change itself can be measured, it is impossible to determine how much change would have occurred anyway, in the absence of the selected element.

There are many measures of technical change, including change in the volume of output, change in the return on limiting resources, change in the income and well-being of producers and other beneficiaries of the production system. However, I am going to concentrate on what I know is a crude, difficult and sometimes inappropriate measure — yield per hectare. My reason for doing this is that even where the first object of producers is to increase the return on limiting resources other than land area (such as labour in periods of peak demand), an increase in output per unit of land area may often help to achieve it. Indeed, over the longer term perspective for the future, the requirements of mankind for food and other agricultural products will largely be met by considerable increases in average yield per hectare.

I shall be particularly concerned with developing countries. They are the scene in which the main agricultural problems of our time, and of the century ahead of us, will have to be resolved.

Let me state one of my conclusions now, in order to foreshadow much that will follow. Much change has occurred in agriculture without benefit of research and extension; and much good research, dedicated extension and competent communication have failed to attain the desired results. With the reasons for all this we shall be concerned later on. At this stage it may be enough to say that it is usually difficult to separate the effects of the knowledge sector, including extension, from many other factors which affect the processes of change.

Experiences of Technical Change in Agriculture

Our experience of agricultural change is substantial, and much of it is well-documented. We were all underdeveloped once: agriculture in the world today is the product of 10 000 years of change. It is only a few hundred years since the yields of crops in Britain were similar to those in sub-Saharan Africa today — around 800 kilograms per hectare of cereals and 4 or 500 kilograms per hectare of pulses. In Britain we have been through a long series of technical changes or revolutions in production methods which have increased the yields of crops and the productivity of the agricultural systems in which they are produced. One needs only to refer to Coke of Norfolk, Jethro Tull and horse hoeing husbandry, Townsend, turnips and the Norfolk four-course rotation, and John Bennett Lawes of Rothamsted, to be reminded of some of the main participants, in the days before extension was invented. During the nineteenth century scientists such as Davy, Gilbert, Liebig and Daniel Hall, who brought about the marriage of science and agriculture in Britain, entered the scene, but the changes were largely led by landowners and farmers. Only relatively recently has the lead passed to persons who are not themselves agricultural producers.

We can see some of the effects of change over less than half a century in the records of agricultural advance in the United Kingdom since 1934–8 (Table 1). At that time nearly one and a quarter million people were active in agriculture, about 5 per cent of the economically effective population. Thirty years later, in 1965, these numbers had decreased to 960 000 (4 per cent of the labour force) and in 1983 the numbers were less than half a million, representing 2 per cent of the labour force. In spite of this continuing decline in the numbers of workers, the output of cereals increased from 4·6 million tons in 1934–8 to 21·5 million tons in 1983, and considerably more in 1984. The area had increased from rather over 2 million hectares to nearly 4 million hectares, but the main source of the increase in output was an increase in average yield from about 2·1 to 5·4 metric tons to the hectare. A good many farmers often harvest as much as ten tons to the hectare.

In these great changes many factors have been involved. Surrounding all has been the policy and practice of successive governments, the

TABLE 1
Cereals and the Agricultural Labour Force in the United Kingdom, 1934–8 to 1983

	1934–8	1965	1983
Output, m tonnes	4·63	13·72	21·50
Area harvested, m ha	2·16	3·65	3·97
Yield, tonnes/ha	2·14	3·75	5·42
cwt/acre	17	30	43
Persons economically active in agriculture, m	1·22	0·96	0·48
Per cent of labour force	5	4	2

general economic climate, diversification and change in the general economy and the advance of industry, science and research, the media, and the status and competence of the producers themselves — including a substantial proportion of experimentally minded gamblers who follow their fancy and will try anything once, or even a few times. Extension has been of many kinds, including extension from farmer to farmer. Several sorts of advisers in the private sector have played a significant part. I do not know how to isolate the contributions of the official extension services from all the other means by which knowledge reaches producers. Many of them nowadays know research workers personally, and deal directly with them.

In spite of all the doom-laden prophecies which we hear on all sides, and of not a few disasters, a broadly similar sequence of change can be seen in the world as a whole. In the period 1934–8 the average yield of cereals in what are now called the developed countries was 1·18 metric tons to the hectare. In the developing countries at that time it was only marginally smaller, at 1·13 tons per hectare. The Second World War accelerated change in the developed countries but caused marked decline in the poorer ones. By 1963 average yields of cereals were about 1·76 metric tons per hectare in the developed countries but only about 1·23 in the poorer countries. But after about 1960, the global green revolution, which had begun in the developed countries during the war, began to move into the developing countries also. In 1983, when the average yield of cereals in the richer countries was 2·55 metric tons to the hectare, the yield in the poorer countries had reached nearly 2·1 metric tons. Changes in harvested area were significant but smaller, and as a result the output of cereals increased over the whole period from 355 to 823 million tons in the richer countries, and from 299 to 844 million tons per year in the developing countries, where in spite of an increase of human numbers by a factor of 2·5, output of cereals per head had increased from 222 kilograms in 1934–8 (followed by a fall to 190 kilograms in the late 1940s) to 243 kilograms per head per year. This may seem a small gain, but it is real, and is, moreover, a great deal better than a loss.

Particularly important changes had occurred in the United States, in Japan (which over the past 50 years has been into, through and out on the other side of its agricultural revolution), and in Western Europe where today we are confronted by very large surpluses of foods of many kinds. In the poorer countries, notable progress has been made in the last twenty years in India, where output of cereals has increased from around 88 million tons per year in the mid 1960s, excluding the two years 1965 and 1966, in which the failure of the monsoon provided the springboard for the subsequent rapid advances which brought output to about 170 million tons in 1984. Over this period yields of all cereals increased by about 600 kilograms to 1550 kilograms per hectare in 1983. (These numbers include rice as paddy, not as milled rice.) Important as these changes have been for India, it is evident that technically India has plenty of headroom and that yields could be substantially further increased. Important progress, largely based on improvements in yield, has also been recorded from Brazil and other countries of Latin America, from Pakistan and from Indonesia and other nations of south and south-east Asia, and above all from the People's Republic of China, which has been no mean participant in the green revolution.

But let us not forget that long before science was invented, producers themselves were responsible for most significant technical changes in agricultural systems in what are now seen as the developing countries. If one asks a casual observer to name the common African crops he will surely mention maize, cassava, sweet potato, groundnuts and pumpkins. In fact, since the Portuguese started to bring the American crops to Africa after about 1500, African producers have adopted, and incorporated into their traditional systems of production, a substantial number of American species — maize, cassava, sweet potato, coco-yam, *Phaseolus* beans, groundnuts, pumpkin and other American cucurbits, tomato, *Capsicum* peppers both sweet and pungent, tobacco, cacao and tetraploid cottons. It is often forgotten that the cacao industry of West Africa was originally established by Africans without any help from science, government or even the foreign private sector.

It may also be of interest to record some parts of the recent history of agriculture in Northern Nigeria, since they illustrate very nicely the relationships between science, extension and production in a developing country. When I first visited Northern Nigeria in the early sixties, the Empire Cotton Growing Corporation had produced a type of cotton which was most acceptable commercially and which would give yields amply sufficient to cover the cost of protection against insects, if it was sown at the start of the rains. Local producers wished to grow cotton in order to obtain additional cash income, but they would not plant the crop until around 6–8 weeks after the beginning of the rains. As a result, the long-season varieties, which were all they could get from the ginneries, gave very small yields. Where it was possible on the experiment station to get over two and a half tonnes of seed cotton per hectare year in and year out, the producers outside the station fence got little more than one eighth of this yield.

The research workers blamed the extension workers for what they regarded as a failure; and when the extension workers managed to demonstrate that this was unjustified, because they had held the prescribed numbers of demonstrations and farmers' meetings, and that plenty of seed was available, the research workers and the extension workers joined together to blame the disappointing outcome on socio-economic factors. This effectively put the whole matter beyond the reach of further investigation because nobody knew what socio-economic factors were, though it was strongly suspected that they were a cover-up for the innate idleness and arrant conservatism of the producers, who were, moreover, supposed by some to be insensitive to civilised economic considerations, to over-value leisure, and consequently to embrace what was called a 'backward sloping response curve'.

Not very long after, socio-economists did indeed begin to work in Northern Nigeria, and by 1968 the answer was plain. The cotton offered by the research workers was in direct competition for labour, at the peak period of the year, with the staple cereals on which the lives and survival of the producers depended. No ordinary producer would allocate so scarce a resource as early season labour to a non-food crop. Only when the food crops were secure — when they had been sown, the gaps had been filled, the weeds had been controlled and a prostrate type of cowpea had been intersown among them — would the producers turn to cotton.

One obvious answer to the dilemma was to breed a cotton adapted to late sowing, so that, though it would not give as large a yield as the early sown long season cotton, it would nevertheless give a great deal more than 250 kilograms per hectare. This was indeed done, but when last I went to Northern Nigeria I found that the cotton breeders had happily returned to the long-season types on the ground that their yields were much better than those of the wretched short-season types which had misguidedly been developed for late sowing.

Soon after my first visit to Northern Nigeria, the all-weather road was completed from Kano to Zaria. Almost at once a very substantial production of soft cane, covering tens of thousands of hectares, sprang up in the bottom lands adjacent to the new road. Rather than use their surplus mid-season labour to grow poor crops of cotton, many farmers and their wives turned to growing cane, for which an immediate market continually flowed past on the trucks which went back and forth along the road. The road had brought the market place to their doorstep; and cash in the hand was better than cotton in the bush. At one blow, they demolished the backward sloping response curve. The purchasers bought the cane for a penny a stick, consumed it and were content, or else they carried it off to the next village and sold it for two pence. 100 per cent profit in 20 minutes is good business even in Nigeria. Other producers turned to growing vegetables, and a very successful business they have made of it, too.

More recently, as the output of cotton, and of groundnuts also, has continued to fall, maize has been developed on a substantial scale in the wetter parts of the region, mostly for sale to feed chickens. As small-scale traditional agriculture has declined with the rise of new economic opportunities elsewhere, Nigerian private-sector investors, mostly merchants, have taken up medium-scale mechanization to produce food to meet the ever increasing urban demand.

These substantial changes have benefitted from substantial investments in physical and managerial infrastructure, but of the successful commodities and techniques only the maize has been significantly aided by research, education or extension. The changes have largely depended on the knowledge, the initiative and the organizing capacity of members of the traditional northern society.

Determinants of Technical Change in Agriculture

In any process of technical change in agriculture, knowledge, old or new, is clearly always a central component; but research and extension have been significant only in some more modern instances. Research and extension usually enter the scene and become effective in association with many other important changes. Modern agriculture is the agriculture of cities. It depends on cities for its market and for its equipment and for much of its science and technical support. Agricultural change in modern times is associated with increases in the size and complexity of economic systems, with industry, with communications and with urbanization. Moreover, it is evident that new knowledge is not always a necessary condition of change, and that it is virtually never a sufficient one. It may sometimes appear to be, in developed societies where the other determinants of change are managed in other ways, so that new knowledge is the principal visible limiting factor. But this is really an illusion: the other determinants are there all right, but they have been well enough managed to be taken for granted. Let us look at what they are.

For this purpose I use a checklist of the principal factors which at different times and places have promoted or opposed technical change in agriculture. It is within the framework set by these principal factors that the knowledge system has to operate and extension has to work. The exegetical text for this part of my remarks consists of three principles enunciated by that distinguished nutritional biochemist Philip Handler, when he was President of the National Academy of Sciences of the United States. He said that it seemed to him that three propositions were sufficient to outline the problems of agricultural development in our time. The first was that we already produce, in the world as a whole, far more food than is necessary for the normal growth, health, activity and reproduction of all the people who live in it. The second was that nowhere in the world does a man go hungry who has money in his pocket; and the third says that nowhere in the world is there a farmer who will produce more than he wants for himself and his family unless someone else gives him something acceptable in exchange.

To these three brilliant truths I have presumed to add a more specious paradox of my own, which says that no animal species known to me is able to increase in numbers for the reason that more of those born are surviving rather than dying in infancy, and at the same time be increasingly short of so critical a resource as food. I feel that there must be some limit to the extent to which even so ingenious a species as ours can have it both ways.

The Five Groups of Factors

I group the factors which determine the nature and rate of change in agriculture and in the rural space into five categories. The first is all embracing and enveloping and tends to dominate all the others: it is the *policy and practice of governments*. In the richer countries governments can afford, at least for a time, to distort policy in ways which do not make economic sense for the nations, even though they may benefit producers and rural people. In the poorer countries, ill-judged government intervention (for example, in maintaining unrealistic exchange rates and prices, incompetent and often corrupt marketing organizations and inefficient state farms) wastes national resources and frequently stifles the production of marketable surpluses. Some of the reasons for this are political, in the pressure on governments to maintain a supply of cheap food for the urban populations who can shoot the President, but some are even deeper-rooted.

All too many of the developing countries entered into independence without a stock of competent people either to direct or to execute national policies. Most of the new nations lacked, and many still lack, both competent politicians and competent civil servants. It is fashionable in our time to jeer at both these classes of persons. But no government can get along without civil servants, competent to advise the political rulers and able to execute their decisions, and no system of government, particularly if it sets out to be representative, can work without a sufficient number of politicial persons competent to ask the right questions, take sound advice, form viable development policies for the rural space in a changing society, and oversee their execution.

Some developing countries had competent civil services at the time of independence: all too many of these have decayed because politicians have not known how to manage them. Very few newly independent countries had a sufficient stock of competent political leaders. In colonial practice, politicians were seldom trained: usually they were put in jail (where, it is true, some of them advanced their education by private study and taught themselves how to do the jobs they now perform with varying degrees of success).

Maybe knowledge services are at least as important for these two classes of people as for agricultural producers and others who live in the countryside.

The second main group of constraints relates to the *volume of effective demand*. It follows from Handler's third principle that no surplus will be produced unless there is an effective demand for it. Many who promote production campaigns seem to believe that it is the duty of the poor farmers, with whose well-being we are all concerned, to deliver free lunches for the benefit of non-rural people. The poor farmers do not agree. To obtain surplus production from the rural areas it is not enough to promote new production options: a satisfactory price for a sufficient volume of produce must be paid. This applies equally to food crops for consumption in the towns, to crops which substitute for imports or provide raw materials for national industries, and to export crops for marketing abroad. Without effective demand at an acceptable price nothing will happen.

The third set of factors relates to what I call the *output delivery system*. This consists largely of physical and managerial infrastructure — roads, bridges, ferries, ports, vehicles, fuel, competent operation, maintenance, markets and market masters who can keep track of quantities and prices, storage, processing, wholesaling and retailing. Only to the extent that the output delivery system is effective will rural people be able to respond to the signals of price and effective demand in the market place in the distant city.

The fourth group consists of *resources for additional production*, and it is perhaps here that some of us begin, like the Ancient Mariner, to recognize some features of our home landscape. These resources include land, water, power and labour, equipment, seed and planting material, fertilizers and agro-chemicals, and credit and many more. Moreover, some of these scarce resources, and particularly labour (with which I bracket time and attention) may be needed for fetching water, preparing food, grinding grain, collecting timber and fuel, and many other necessary activities in the complex life system of most rural people in developing countries, in which it is often neither socially nor economically feasible to isolate a separate agricultural sector for analysis.

Only in so far as these four groups of factors are in a reasonably satisfactory state is it possible to consider the fifth group, the *technical methods for increasing output*. Put another way, only to the extent that the first four groups of factors are in a satisfactory state can there be a market for knowledge services and an effective function for extension. Conversely, where enough of these factors are in a satisfactory state, change can occur using the existing knowledge of producers, without benefit of research, education or extension.

THE ROLE OF EXTENSION

It is in the intermediate area between these conditions, where producers see that certain forms of technical change are likely to be to their advantage, but need to learn how to use them, that they will seek the help of knowledge services. In these circumstances, the returns to extension may indeed be very substantial even though they may still be very difficult to measure.

But, in fact, the role of extension is far wider and far more responsible than the communication of knowledge. Its fuller task is to comprehend the social, economic and technical environment of the producers, and their objectives, resources, existing methods and difficulties; to represent their needs to the administrative apparatus as well as to the rest of the knowledge system; and to help convey the products of that system, and options derived from them, to producers. The extension worker then becomes an active participant, close to the front line, in the processes of change in agriculture and in the rural space.

These tasks contain a great deal more than messages coming down from on high and feedback coming back from the producers. This common formulation seems to me to do no more than give a superficial air of respectability to what is still essentially a top-down process.

In this connection, I feel bound to venture some comments on the Training and Visit system which the World Bank has so vigorously embraced. The great merit of the T & V system seems to me to be the technical, logistic and moral support it offers to the extension workers. Though this may be vigorously denied, its main defect seems to be that it is not the producers who decide what they need to know but the managers and leaders of extension in the office who decide what producers should do, what they need to know, and when they are to be .told about it. I believe that, except in some emergency situations, the producers should come to share the leadership and have a significant, perhaps the most significant, voice in determining what they need to know and when. After all, it is they, and not those placed in authority over them, who commit resources, do the work and take the risks.

No doubt, at a price, the T & V system is better than the disorganized, unsupported and under-funded pretence that must often have preceded it, but I believe that a producer-led system, or a system in which the producers are at least seen as partners, and in which extension serves rather than commands the producer, must often do better at smaller cost.

The conceptions I have outlined suggest a complex and intellectually attractive and rewarding task for extension.

3

Extension and the Development of Human Resources: the Other Tradition in Extension Education

NIELS RÖLING

Department of Extension Education, Agricultural University, Wageningen, The Netherlands

INTRODUCTION

The period of twenty years which my career in rural extension spans should be sufficient to foster detachment in man and institution. Even where both continue in hot pursuit of seemingly crucial issues, however, the kaleidoscopic change over two decades imposes some relativism, whether one likes it or not. I am going to turn this state of affairs to my advantage by attempting to look at extension science in the same manner as extension science looks at extension practice, or for that matter, as extension practice looks at farming and as farming looks at resources and opportunities. Fig. 1 illustrates this.

Resources ←--- Farming ←--- Extension ←--- Extension ←--- This
opportunities   practice   science   presentation

Fig. 1. The chain of critical reflection and discourse in rural extension.

One can look at Fig. 1 as a chain of intervention: activities in the realm of farming can be seen as the outcome of intervention by extension practitioners. Likewise, the activities of extension practitioners can be seen as the outcome of interventions by extension scientists.

Figure 2 elaborates on the first figure by indicating the main areas of discourse in our field of interest. I feel it is important to make the distinction between 'gets' and 'wants', between actual and normative, and between 'do's' and 'should do's'. If there is one area in which actual

	Resources opportunities	Farming	Extension practice	Extension science	This presentation
'GETS' 'DO's' (actual outcomes, empiry)	1	3	5	7	9
'WANTS' 'SHOULD DO's' (normative images, intentionality, *Leitbild*)	2	4	6	8	10

Fig. 2. Areas of discourse in rural extension.

and normative are mixed up, it is in extension. We even define extension not in terms of what it is, but more often in terms of what it should be. In other words, we reify cell 6 into cell 5.

What is more, we often act as if the intentionalities (the evenly numbered cells) are consistent across the chain, or we try to make them more consistent by acting as if they are. Thus, some have made contributions to extension science in the area of small farmer development in cell 7, which implies a desire for equitable development in cell 6, while the focus in cell 4 is, in fact, more often on national food security or on maintaining the export position which is consistent with a focus on progressive farmers in cell 5.

Such inconsistency does not lead to an effective chain of intervention. This can be deduced from the fact that the focus on small farmers and equity in publications such as Ascroft *et al.* (1973) and Röling *et al.* (1976), and in textbooks such as Adams (1982) and the FAO Manual re-edited by Swanson (1984), does not seem to have led to much effect in current practice, while the T & V system, which is much more consistent with the present content of cells 4 and 6, has spread very rapidly. Figure 3 illustrates the mechanism at work. It shows the 'flow of intentionality' in the field of rural extension.

For example, extension scientists hold courses (cell 7) in order to affect what extension workers intend to do (cell 6), which hopefully affects what they actually do (cell 5). What extension workers do affects what farmers intend to do, and so forth. We are, in short, in the business of affecting intentionalities. Our instruments, such as courses, books,

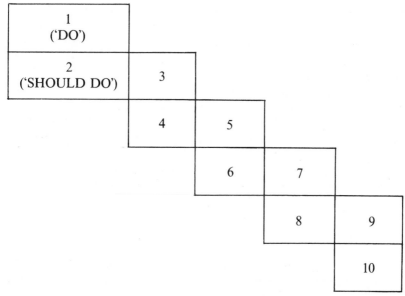

Fig. 3. The flow of intentionality in the field of rural extension.

demonstrations and so forth, consist of communication. We can only *derive* power, be it from the context or from our affiliation with power holders. We do not hold or exercise it directly. That is, our interventions, be they from extension worker on farmer, from extension scientist on extension worker, can only be effective through voluntary change. A coincidence of 'WANTS' and 'SHOULD DO's' is a condition for effectiveness. This implies some user control, or at least a sound knowledge of the potential for voluntary change. This applies not only in the case of the relationship between extension practice and farming, but also between extension science and extension practice (and between papers such as this and extension science).

I must confess I have often tried hard to make people want what they should do — according to me. I am not saying that I like our impotency. I, too, like things my own way rather than anybody else's. But if we forget the logic of our instrument we achieve nothing. We know this all too well in the case of extension. We know that if we use extension as a policy instrument without regard to farmers' intentions we either alienate the farmers or our field workers. In fact, we know that the present drive to economize and make extension more accountable has increased the friction between these officials who are close to the top and the field workers who face the farmers. This friction leads to fascinating phenomena in the bureaucracy. In extension science it is the same. If we try to teach extension workers, we are considered to be 'abstract', 'academic', or impractical.

But all is not hopeless. In between the two extremes of effect but no change and change but no effect there exists a small amount of scope for effective change through persuasion and knowledge utilization. Here lies the potential for voluntary change and for affecting 'WANTS' by 'SHOULD DO's'.

This, then, is the context within which I would like to place this paper. It falls squarely in cell 9 and is subject to all the pitfalls I have described.

TWO TRADITIONS IN EXTENSION

The theme of this paper, and that of Professor Bunting on technical change, reflect the main traditions in extension education, which I shall call: 1. technical innovation (TI), and 2. human resource development (HRD). I shall discuss the relationship between these two traditions before dealing with HRD in greater detail, since I believe HRD cannot be looked at in isolation from TI.

TI is a strong tradition in extension. Key words include diffusion of innovations, adoption, farming systems research, research/extension linkages, appropriate technology, results demonstrations, T & V, agricultural knowledge systems, the commodity approach, and others. This list shows how centrally TI stands in our concerns. What is more, I submit that most of the world's extension workers and agencies are engaged in pure TI. The farmer is an instrument through which resource use is made more efficient and productive. Extension is financed by tax revenues for purposes of TI. It is in the first place an instrument to make the production of food, raw materials and export commodities as effective and efficient as possible.

HRD is an entirely different tradition. Key words here are community development, institution building, emancipation, leadership development, normative/re-educative strategies, mobilization, organization, developing delivery systems, and so forth. The focus is not on developing natural resources through people, but on rural people themselves and on the social systems in which they function.

For example, most extension work organized according to T & V focuses on TI. But reorganizing an existing extension service according to T & V principles is a new way of organizing people and can be considered as HRD.

The example shows that TI and HRD should be seen as two dimensions of extension work which mutually reinforce each other. But, in actual practice one often finds them functioning in opposition to each other and with different supporters. Each of them then develops in isolation which leads to a lack of mutual correction. Whereas TI is typically carried out by technically trained government employees who do not feel comfortable with HRD, HRD is often carried out by people with a social science training and with sometimes little regard for the limitations imposed on human intentionality by the nature of things. In many countries HRD is the work of non-government organizations (NGOs).

A typical example of the conflict between the two approaches is a Dutch pilot project in consumer education for the urban poor. The project works with five pairs, each consisting of a social worker and a home economist. The social workers seek to organize the poor so that they can improve their opportunities. The home economists seek to provide technical information to help people make better use of the opportunities they have. Though the two could be seen as mutually reinforcing in theory, in practice the pairs have been in intense conflict and the pilot project has not been very effective so far.

The two traditions differ in terms of social acceptability. TI is based on hard criteria and science. It is spurred on by competition, the need for food and other necessities. Few people quarrel with TI, except over some of its consequences. One could say that TI is a mechanism which plays the same role in society as genetic evolution does in nature; the process of TI is, in fact, very similar to evolution.

HRD is much less socially acceptable. HRD goals are less easy to defend on the hard criteria of efficiency, effectiveness and survival. Many HRD goals such as equity, emancipation or participation are not politically neutral in that they seek to affect the distribution of power and resources. Thus, whereas TI is political in its consequences, HRD is often political in its intentions.

The two traditions in rural extension reflect the situation in society at large. Spurred on by the conflict between capitalist and communist systems, by economic competition, and by each individual's search for survival, comfort and control over the environment, science and science-based technology are predominant in shaping our values and social systems. We have allowed them to become deterministic of our futures.

The two traditions in rural extension also reflect important distinctions made in modern philosophy. An example is the 1984 BBC Reith Lecture Series (Searle, 1984) in which mind and brain, behaviour and matter, intentionality and scientific law, social science and physical science, and free will and determinism are juxtaposed and discussed in terms of their interface in human action, in knowledge and control.

Looked at in this way, the two traditions in rural extension are linked to the most fundamental of human concerns. They reflect our uneasiness with our lack of a self-chosen *Leitbild* and with the powerlessness with which we are carried along the rapids of TI towards consequences which make us shudder. In this respect, extension is a fascinating subject because it deals explicitly with such issues in the more limited and practical field of agriculture and rural development. Within this microcosm one is constantly confronted with the need for shaping society and with the limited scope available for doing so.

TWO DIMENSIONS OF PLANNED SOCIAL CHANGE

Brakel (1985), a former personnel officer of an oil refinery with 6000 employees, has observed that changing people and changing tasks are two dimensions of effective change. Developing only one leads to ineffectiveness. He uses a skating analogy: if one only develops people and not tasks, one skates on one leg and turns around in circles. The same happens if only tasks are developed. One only goes forward if one uses both skates.

TI and HRD in extension should similarly be considered as aspects of the same thing. The opportunities offered by TI should be fitted into efforts to shape society according to deliberately chosen *Leitbilds*. On the other hand, opportunities created by TI should affect the goals chosen. Two examples will illustrate how TI and HRD can work together in practice. One comes from Bangladesh and the other from Canada.

In an increasing number of developing nations such as Bangladesh, NGOs have initiated programmes of rural poverty alleviation which may involve thousands of people. A typical example is the Bangladesh Rural Academy (BRAC) (Bouman, 1984) which has effectively provided start-up capital to hundreds of small groups of rural poor. These groups have been deliberately created and trained to act as platforms for discussion, participation and countervailing power which allow conscious decision-making according to futures which have been deliberately chosen and discussed. Such efforts as those of the BRAC are beginning to show an impact in areas which seemed impossible to lift out of their misery. They also reflect a mixture of TI and HRD. In fact, they consist of an essential mix of ingredients which one finds again and again in the literature (Colin, 1978; Jiggins, 1983a; Röling and De Zeeuw, 1983):

1. Mobilization ⎫
2. Organization ⎬ of rural people through
3. Training   ⎭ HRD by NGOs
4. Tangible opportunities provided through TI by technical agencies

If such approaches limit themselves to HRD, they soon collapse. But a limitation to TI alone does not allow the rural poor to grasp the available opportunities. A carefully monitored field experiment by Verhagen (1984) has clearly demonstrated this.

The second example is the Advanced Agricultural Leadership Program organized in Ontario by the Foundation for Rural Living, the Ontario Federation of Agriculture, the Ontario Ministry of Agriculture and Food, and the University of Guelph. The programme seeks to train some 30 men and women during 50 days spread out over a two-year period and which includes twelve seminars and a study trip abroad. The men and women chosen are between 25 and 40 years of age and are in the early stages of leadership careers in agriculture, agri-business, policy-making, civil service or politics. The programme covers policy- and decision-making, economics, agriculture's impact on the environment, fiscal and monetary policies, commodity trading, taxation, consumer issues, land use, media and trends in agriculture and society.

Such a leadership programme obviously aims at HRD and focuses on creating a capacity to make deliberate choices in a society characterized by a sophisticated technology and rapid innovation. The programme also reflects a new world concern for community development, civics, and institution building.

TI's USUAL SPLENDID ISOLATION

Though examples do exist where TI and HRD occur together, their scarcity highlights the extent to which attention in agricultural development and extension has been absorbed almost exclusively by TI. There are some good reasons for this.

In the first place, the threat of mass hunger and rapidly growing populations have led to an urgent search for technical solutions which could avert disaster. In fact, a simple calculus of available arable land and mouths to feed leads to the conclusion that TI involving high yielding varieties, the use of fertilizers and other agricultural chemicals is already lagging behind.

In the second place, maintaining one's export position in the fierce competition for foreign markets leads to a prime concern for efficiency of production. If one loses one's competitive edge, others will take over. Hence, there is a scramble to make TI as effective as possible; concern for the consequences of TI on landscape, environment or rural society are swept away by the export argument.

In the third place, the very process of TI leads to its predominance. Farmers are eager to improve their incomes. New technologies allow them to produce more or more efficiently. The early ones use these technologies at existing price levels and make a windfall profit (Rogers, 1983). Soon, however, others follow. Total production increases and prices drop (or remain behind relative to those of industrial products). Those farmers who have not innovated are forced to do so lest their incomes drop; those who cannot will eventually drop out of agriculture altogether. In this sense, a new and more efficient technology closely resembles a superior gene in the evolution of a species.

Given such basic propellants behind TI, the glamour of the IARCs and the lure of World Bank funds supporting TI are only secondary phenomena which support my contention that most extension and extension science is now purely focused on TI.

In fact, the main issue in present-day extension science is not a juxtaposition of HRD and TI, but *within* the TI tradition, the need to at least anchor the development of technology in the rural society through participation, rapid rural appraisal, farming systems research, on-farm and with-farmer experimentation and evaluation. The need for such anchoring of technology generation, transformation and utilization in rural society is raised in extension science in the face of the predominant focus in current practice on the transfer of technology (Fig. 4) (Chambers and Ghildyal, 1985).

Extension scientists and others, however, are making an effort to introduce a model which is more farmer based (Fig. 5). However necessary the effort to introduce the farmer-based approach, it should

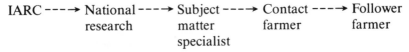

Fig. 4. The transfer of technology model.

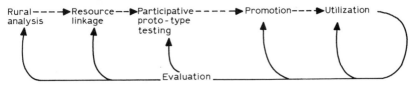

Fig. 5. Farmer-based technology development, transformation and utilization.

be noted that we still stand squarely within the TI tradition. The difference between the two models is that the former says: 'have technology, look for farmers who can utilize it', while the latter says: 'have category of farmers with technical problem, look for technology which can solve it'. But neither of them deals with HRD. Extension science, by aiding and abetting the single-minded focus on TI, abrogates its responsibility. It will not be capable of developing answers to societal problems which are already in the offing for those who are willing to see them.

One of the surprising characteristics of our era of rapid rural change is the lack of attention to the rural societies which are implicated in the present processes of change. We worry about food availability, but that is about it. We pay little attention to deliberately choosing futures. We give all our effort to stimulating TI and accept whatever consequences that might bring us. Thus, we all know that most developing countries will have a sizeable majority of rural people in the years to come. We know that we should avoid a situation in which rural masses become so impoverished that they are forced to migrate to urban areas. Yet, we have no answer to the question of how these rural people are to survive in the face of rapidly decreasing prices of agricultural products due to rapid technical development in high potential areas.

In Europe the crisis is upon us. The cost of price subsidies and of maintaining agricultural surpluses has become unbearable. The question now is whether to let TI take its course to its logical final state of one cow in one polder which produces all our requirements, or whether to pay farmers a regular income so as to maintain farmer employment; the Commission of the European Communities appears to be preparing proposals to do the latter.

Three types of costs resulting from an exclusive focus on TI may be enumerated: costs internalized in the households, (e.g. female malnutrition); direct or indirect external costs (e.g. labour pushed out of agriculture to subsist in the margins of urban informal economies); and costs postponed into the future (e.g. environmental costs). Other examples of the costs of an exclusive focus on TI include the need for price subsidies, rural unemployment, the lack of effective support systems for technological development, or the technocratic nature of the available support systems, which has been called the Technical–Administrative Task Environment (or TATE).

In some cases, the costs are becoming painfully apparent. In the Netherlands a direct link between the spreading of large quantities of liquid manure produced by the intensive bio-industry and acid rain has been clearly established, as has been the pollution of underground drinking water supplies by dumping such manure. Any expansion of pig and poultry farming has now been prohibited in the affected areas.

In the Philippines in 1985, farmers demonstrated against IRRI. They see it as a political instrument which has made them dependent upon fertilizer and delivered them to the extractive practices of industrialists. Given rising fertilizer prices and falling prices for agricultural products, they are finding it impossible to utilize the high input technology. For the first time in more than a decade, the Philippines had to import rice in 1984.

The time seems to have come, not for a complete reversal of the trend, not for a complete pendulum swing to social science, HRD and 'soft stuff' in general, but for a more balanced consideration of agricultural development and for attention to the deliberate choices which are necessary for shaping desirable social systems through HRD.

Thus, I wish to consider some aspects of HRD in a preliminary attempt to create some system in our thinking about it.

AN ATTEMPT TO SYSTEMATIZE HRD CONCERNS

I shall develop a systematization of HRD in extension* by focusing on the costs of TI; in fact, dealing with HRD as if it were a method to correct some of the main problems caused by TI. This is a limited focus, but I believe that an indication of ways by which some of the costs of TI

*The present paper specifically excludes formal agricultural education as an instrument of HRD.

can be redressed is of the highest priority. I shall enumerate some of the HRD approaches which seem suitable or have been developed for each main cost.

Dependency on External Sources of Information

A focus on TI can easily lead to over-reliance by farmers on external sources of information for decision-making and to reduced control over farm development. Due to TI's not infrequent bias in favour of high-tech solutions, and lack of attention to local opportunity, over-reliance on external sources of its 'recipe' information can lead to decisions which threaten long-term survival. To redress this situation, HRD seeks to develop management capacity by developing feedback loops, information processing, economic decision making and policy analysis. A typical example is a Queensland programme to train farmers in systematic storage and retrieval of written information.

Ineffective Institutional Support for TI

The lack of institutional support for TI (e.g. the absence of 'delivery systems') is the one area of HRD which finds some favour in the present TI-oriented climate. One recent publication (Cernea, 1984) draws attention to the need for social engineering to create the institutions necessary for effective TI. Likewise, the development of T & V systems and research/extension linkages has been stressed as an essential condition for effective TI. A special international service to help develop national agricultural research institutions (ISNAR) has been developed as part of the CGIAR system.

Technocracy Hinders Adjustment to Rural Conditions

In many countries, the one-sided and technocratic development of *structures d'intervention,* as the French so aptly call them, has led to the promotion of inappropriate technology, to the formulation of inadequate agricultural policies, to insensitivities of services to rural conditions and to frustration among farmers. In these conditions, HRD focuses on developing countervailing power or user control on the part of utilizers and utilizer systems. In fact, such highly successful TI systems as the Land Grant Colleges in the US and the Dutch government's 'Service for Enterprise Development' are characterized by a high degree of user control (Rogers *et al.*, 1976). In Kenya, Leonard (1977) has named the influence of farmers on the government agencies the 'squawk factor' and noted its importance in adjusting the agencies to farmer conditions. Of course, in all countries where user control is present, it is being used to adjust the agencies in favour of influential and high access farmers. The HRD challenge is to develop the 'squawk capacity' in other categories of farmers.

Deterioration of Terms of Trade *vis-à-vis* Industry

One of the costs of TI is the deterioration of agricultural prices relative to those of industrial products and services. In the Netherlands, an egg now costs about the same as in the 1950s, which is, of course, a boon for consumers. What we are talking about, however, is that it is a consequence of TI which might reach a limit in terms of its societal benefit. HRD efforts to redress such costs include the development of farmers' unions which can influence national policy-making, leadership development (e.g. the Canadian example given earlier), civics training and so on. Such HRD contributes to strengthening the agricultural sector in defending its interests against those of other sectors in society. It seems preferable to have a highly developed agricultural sector which is capable of bargaining for its position, than to be faced with either an impoverished agricultural sector, or wildcat actions and the need to import food, as is shown by the Filippino example mentioned earlier.

Costs Postponed into the Future

Present agriculture and its focus on TI is incurring environmental costs which cannot be sustained indefinitely. Our present food production is based on non-renewable energy resources. The same goes for our plans for coping with population growth (e.g. in India). To my knowledge, alternatives do not exist, nor is there serious consideration of alternative systems. More ecological approaches are presently sponsored by fringe organizations and individuals who are not taken seriously by our leading agronomists and development institutions. Moreover, that is not the only aspect. The present agricultural technology used on the former prairies in the US and Canada, or in the Amazon basin, is known to be non-sustainable and necessarily will reach its end in a calculable number of years. HRD to redress such societal costs would need to focus on the development of institutions which can measure the consequences and are capable of affecting public opinion sufficiently to make an impact.

Expulsion of Rural People Out of Agriculture

One of the best publicized costs of TI is the 'marginalization of the rural poor'. Of course, this terminology is dated and has lost its appeal as the ineffectual leftist intellectuals have been stifled and new realism, Thatcherism, Reaganism, or whatever the local national variety of the phenomenon may be called, takes over. The falling out of favour of terms and intellectuals, however, should not blind us to the very real process by which TI leads to the expulsion of smaller and less efficient farmers from agriculture. In the Netherlands, the annual average has been 2 per cent since the 1960s and fully one third of Dutch farmers are still to go. We may, of course, ignore the process. My contention is that we cannot afford to do so, especially in developing countries where the costs of income support cannot be carried and where alternative employment in industry and services does not seem to be forthcoming rapidly. In these conditions, a considerable effort in HRD seems necessary which focuses on emancipation, that is mobilization, organization, training and appropriate opportunity development to create rural livelihoods for masses of small farmers. I have already given an example from Bangladesh to indicate that such HRD is not only possible but is proceeding on quite a large scale. I believe that the international development agencies and governments should support such HRD on a much larger scale than is presently the case.

Depression of Rural Areas
Where categories of rural people lose out in the TI process, the same can be said for rural areas which have a low potential for utilizing high input technology. They tend to be depressed through much the same processes as lead to the expulsion of individual families out of agriculture. HRD, to redress this cost, would focus on community development, institution building, infrastructural development and on mobilizing political power so as to improve representation of such areas in the national arena.

This concludes the enumeration of HRD approaches to redress costs of TI. Of course, the enumeration could have been much more sophisticated, containing many more examples and elaborating methodological detail. That is not my purpose. Besides, I am not certain whether we, extension scientists, are capable of providing such great detail on substance and method. Our focus on TI has left its mark on our body of knowledge. Our theories and concepts reflect TI and are not appropriate for human resource development.

In the enumeration of HRD approaches I have not mentioned the word 'participation'. This has been deliberate. 'Participation' is an imprecise term which leaves room for a great many interpretations. It is, therefore, ideal in any discourse between a social scientist and a technical scientist. But in a forum of extension professionals such concepts are best avoided. Each of the different forms of HRD which I have described has been called 'participation' at one time or other, quite apart from all those instances where participation has been used in a purely instrumental fashion to further TI.

CONCLUSION

The above consideration of various forms of HRD has naïvely disregarded political realities. It is right to ask: and who is going to take the decision to fund such HRD? Where are the sources of political support for such actions which are presently in no one's short-term interest?

The point is, however, that there are two sources of influence for societal change. One is power and political influence. But the other is knowledge, proof and conviction. What is more, political power can be generated around a cause and around sound knowledge. Public opinion can be mobilized around issues which are sound and based on fact.

Professionalism in our field is based on the ability to find the narrow scope for effective change in between the extremes of change without effect and effect without change. As extension scientists, who know and can see the problems with TI, we have a responsibility to use our professionalism to seek change and support for a more balanced mix of TI and HRD. This means that extension as a professional field and as a body of knowledge should begin to carry out the research, write the textbooks, build the curricula and otherwise develop its body of instruments to redress the present imbalance. In this quest, of course, it should avoid over-reaction. HRD is not neutral. It can also have enormous social costs.

ACKNOWLEDGEMENTS

Helpful comments from the following persons are gratefully acknowledged: A. van den Ban, L. Fresco, F. Heymann, B. Huizinga, J. Jiggins and M. J. Rolls.

4

The Economics of Extension

ROBERT E. EVENSON

Department of Economics, Yale University, New Haven, Connecticut, USA

INTRODUCTION

Extension services whether provided by private firms or public agencies must produce economic benefits sufficient to justify their costs. The benefits must generally take the form of increased efficiency of farm production. Extension programmes enable increased efficiency both through the facilitation of the adaption and use of improved technology and through advising better management of farm resources.

Some extension programmes do not produce benefits sufficient to justify their costs. This situation can occur when extension personnel have low levels of competence, when good substitute institutions exist or when little new relevant technology is available to extend. Economists have undertaken a number of studies of extension programme impacts in recent years. In addition, a few economic studies have attempted to investigate the factors determining investment in extension systems. These studies do not provide full guidelines showing how to design economically efficient extension programmes in different countries, but they do provide some indications of the factors associated with efficient programmes. In this paper I will review the studies of extension programme impact on productivity, farm incomes and employment and the value of farm assets. I will also review the studies of political and economic factors that influence investment in public sector extension programmes. Before turning to these studies, however, it will be useful to discuss the interrelationships between extension activities and other programmes that form part of the 'portfolio' of productivity enhancement activities.

CONCEPTUAL ISSUES: PRODUCTIVITY AND TECHNOLOGY

At any given time, the production on a given farm can be characterized as in Fig. 1.

This figure identifies five 'yield gaps' which provide potential for pay-offs from investment in research, extension and market improvement

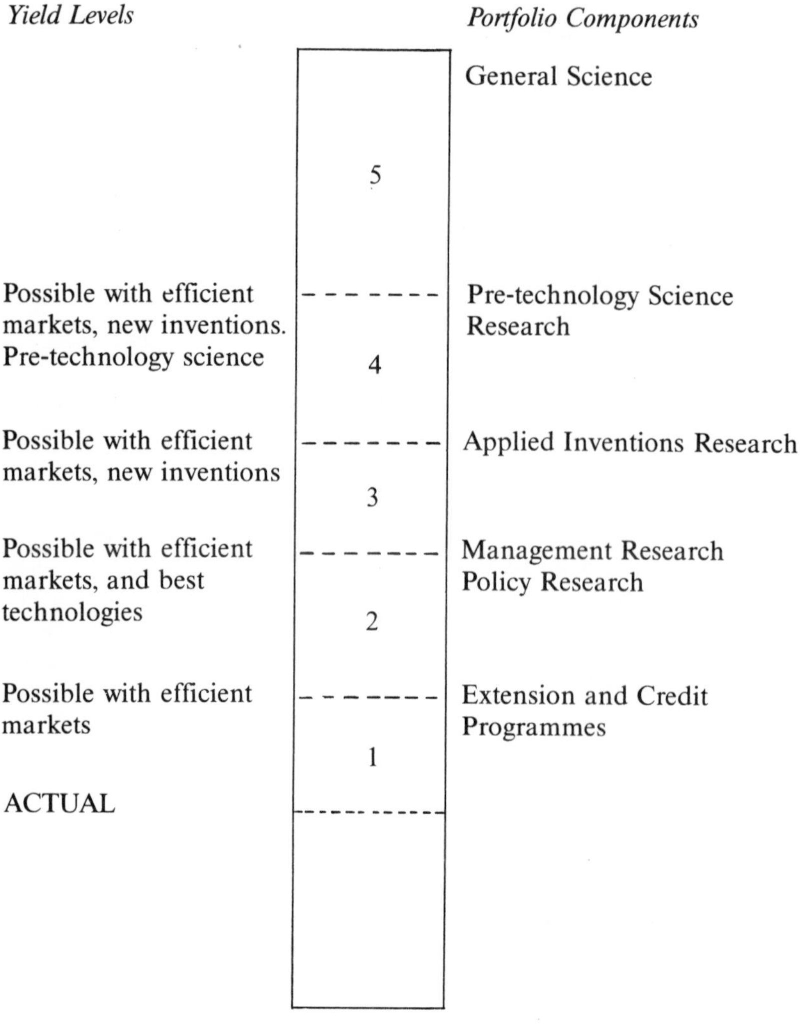

Yield Levels   *Portfolio Components*

General Science

5

Possible with efficient   – – – – – –   Pre-technology Science
markets, new inventions.   Research
Pre-technology science   4

Possible with efficient   – – – – – –   Applied Inventions Research
markets, new inventions
                                 3

Possible with efficient   – – – – – –   Management Research
markets, and best   Policy Research
technologies   2

Possible with efficient   – – – – – –   Extension and Credit
markets   Programmes
                                 1

ACTUAL   – – – – – – – –

Fig. 1

activities (especially credit). It is important to note that these gaps will not remain constant over time. Programmes of extension and market improvement can reduce gap 1, the gap between actual management and economically optimal management with the same technology. These programmes can also reduce gap 2, the gap between optimal management with actual technology and optimal management with best available technology. Farming systems research, research on farm management and most agronomic research is directed towards the reduction of gaps 1 and 2. Gap 3, the gap between best available technology and 'new' technology can only be closed by research directed towards the development of new technology *suited to this farm.*

It is important to note that as progress is made in closing gaps 1 and 2, these gaps become smaller and the potential for further pay-off to the programmes designed to improve efficiency and extend best technology will become 'exhausted'. On the other hand, the closing of gap 3 by inventing new technology initially 'opens up' gaps 1 and 2 and thus provides potential for productive investments in extension and farm management and farming systems research. Another way of expressing this is that if gap 3 is not being closed, i.e., no new technology is being developed (or imported), extension and management research will produce gains that are either temporary or 'once-for-all' (they will be temporary if farmers revert to old practices once the programmes are removed). Gains from new technology, on the other hand, can be continuous.

The exhaustion principle applies to invention research as well, because a *science base* underlies the potential for inventions. If the state of scientific knowledge, the genetic pools, the state of measurement and selection methods is constant, applied inventive research will also become exhausted. Pre-technology science research, i.e., research directed towards improving the science base underlying agricultural invention in the public or private sector, can open up new potential by closing gap 4. This opens up gap 3 and 'refuels' gap 3. When invention takes place, this opens up gap 2 and refuels the management research and extension potentials.

We thus have a kind of 'hierarchical dependence', where upstream components refuel downstream components.

Now consider the technology 'portfolio' of investments that a country can make. Countries differ greatly in their portfolios at different stages of development. Some countries have practically no investment in technology and rely on other sources for growth. Historically, countries have generally invested upstream only when downstream sources are perceived to be exhausted and hence high cost (or low pay-off) sources of growth. Because of the time lag in the recognition of exhaustion, they have generally under-invested in the upstream components of the portfolio. This is why we observe high measured returns to agricultural research.

Optimal investment for a given region or country will depend on the state of exhaustion (or the size of the gaps) of potential, and on the prospects for 'importing' the relevant product from other regions and countries or International Agricultural Research Centres (IARCs). This importing or transfer process is critical. If a region could rely on inventions being produced outside the region, i.e., research organizations producing a flow of new inventions that were cost reducing for that region, then its optimal portfolio might involve only gap 1 reducing activities — extension and market improvement programmes. Many countries in the early stages of development have based their development strategies on this principle. In the 1950s, many countries estimated gap 1 to be large and invested heavily in extension and other rural development projects. By and large, this portfolio strategy proved to be a low pay-off strategy.

It was a low pay-off strategy for two reasons: first, the skill level of many early extension services was insufficient to enable the screening and testing of technology from outside the region. Second, the transferability of technology was more apparent than real. In actuality, technology suited to region A (in the sense that it enables minimum cost production) is seldom suited to region B if economic, soil and climatic factors are different in B. For much technology (e.g. corn varietal technology) surprisingly small variations in soil and climatic conditions inhibit or block transfer.

When countries realized that new technology would not 'spill-in' from abroad, they began to consider upstream components in this portfolio strategy. Twenty years ago, before the advent of 'Farming Systems Research', they usually added applied commodity research programmes. Typically, these had another genetic improvement component (i.e., plant breeding). Thus, they tackled the reduction of both gaps 2 and 3. Today, with the advent of the IARCs and farming systems and 'on-farm' research programmes, some countries are attempting to 'get by' with a portfolio based on the reduction of gaps 1 and 2 and close association with the IARCs. They are not building strong gap 3 reduction programmes. Since these strategies are relatively new, we have little evidence yet as to their pay-off.

These trends in investments can be seen in the international data discussed in the next section of this paper.

SUMMARY OF NATIONAL AND INTERNATIONAL INVESTMENT IN AGRICULTURAL RESEARCH AND EXTENSION

National investment in agricultural research and extension programmes has grown at an impressive rate in the past 25 years. Tables 1 and 2 summarize this investment. It may be seen that, in 1980 constant dollars, research spending in developing countries increased from 1959 to 1980 by a multiple of 5·8 in Latin America, 6·9 in Asia, and 3·6 in Africa. The comparable spending multiples for extension investment were 6·4 for Latin America, 3·5 for Asia, and 2·2 for Africa. Scientist man-year (SMY) multiples were 6·0 for Latin America, 4·1 for Asia, 4·2 for Africa, and for extension workers the multiples were 6·8 for Latin America, 1·8 for Asia, 2·9 for Africa.

Table 3 shows how 'spending intensities' for research and extension, i.e., spending as a percentage of the domestic value of agricultural product (GDP), have changed from 1959 to 1980. These data show that, in 1959, the low-income and middle-income developing countries were approximately twice as spending intensive for extension as for research. The reverse was true for the industrialized countries. The rapid growth in spending intensities for research from 1959 to 1980, combined with little or no growth in extension intensities in the 1970s, produced roughly equal spending intensities by 1980 for research and extension in most developing countries.

Table 4 provides comparable data for 'manpower intensities' (i.e., ratios of manpower to GDP). For research the same general pattern reflected in spending intensities is reflected in the manpower intensities. Because spending per SMY is lower in developing countries they fare better by this measure and the difference between the low-income and industrialized countries is much reduced.

For extension, the picture is quite different. By 1959 low-income developing countries had attained very high extension manpower intensities — five to seven times greater than those attained in industrialized countries. By 1980, with a slight decline in these intensities for industrialized countries, the difference was even greater. Middle-income and semi-industrialized countries also increased their extension intensities.

TABLE 1
Agricultural Research Expenditures and Manpower

Region/Sub-region	Expenditures (,000 in constant 1980 US $)			Manpower (Scientist man-years)		
	1959	1970	1980	1959	1970	1980
Western Europe	274 984	918 634	1 489 588	6 251	12 547	19 540
Northern Europe	94 718	230 135	409 527	1 818	4 409	8 027
Central Europe	141 054	563 334	871 233	2 888	5 721	8 827
Southern Europe	39 212	125 165	208 828	1 545	2 417	2 686
Eastern Europe and USSR	568 284	1 282 212	1 492 783	17 701	43 709	51 614
Eastern Europe	195 896	436 094	553 400	5 701	16 009	20 220
USSR	372 388	846 118	939 383	12 000	27 700	31 394
North America and Oceania	760 466	1 485 043	1 722 390	8 449	11 683	13 607
North America	668 889	1 221 006	1 335 584	6 690	8 575	10 305
Oceania	91 577	264 037	386 806	1 759	3 113	3 302

Latin America	79 556	216 018	462 631	1 425	4 880	8 534
Temperate South America	31 088	57 119	80 247	364	1 022	1 527
Tropical South America	34 792	128 958	269 443	570	2 698	4 840
Caribbean and Central America	13 676	29 941	112 941	491	1 160	2 167
Africa	119 149	251 572	424 757	1 919	3 849	8 088
North Africa	20 789	49 703	62 037	590	1 122	2 340
West Africa	44 333	91 899	205 737	412	952	2 466
East Africa	12 740	49 218	75 156	221	684	1 632
Southern Africa	41 287	60 752	81 827	696	1 091	1 650
Asia	261 114	1 205 116	1 797 894	11 418	31 837	46 656
West Asia	24 427	70 676	125 465	457	1 606	2 329
South Asia	32 024	72 573	190 931	1 433	2 569	5 691
Southeast Asia	9 028	37 405	103 249	441	1 692	4 102
East Asia	141 469	521 971	734 694	7 837	13 720	17 262
China	54 166	502 491	643 555	1 250	12 250	17 272
WORLD TOTAL	2 063 553	5 358 595	7 390 043	47 163	108 510	148 039

Sources: Boyce and Evenson (1975), Judd *et al.* (1983).

TABLE 2
Agricultural Extension Expenditures and Manpower

Region/Sub-region	Expenditures (,000 in constant 1980 US $)			Manpower (workers)		
	1959	1970	1980	1959	1970	1980
Western Europe	234 015	457 675	514 305	15 988	24 388	27 881
Northern Europe	112 983	187 144	201 366	4 793	5 638	6 241
Central Europe	103 082	199 191	236 834	7 865	13 046	14 421
Southern Europe	17 950	71 340	76 105	3 330	5 704	7 219
Eastern Europe and USSR	367 329	562 935	750 301	29 000	43 000	55 000
Eastern Europe	126 624	191 460	278 149	9 340	15 749	21 546
USSR	240 705	371 475	472 152	19 660	27 251	33 454
North America and Oceania	383 358	601 950	760 155	13 580	15 113	14 966
North America	332 892	511 883	634 201	11 500	12 550	12 235
Oceania	50 466	90 067	125 954	2 080	2 563	2 731

Latin America	61 451	205 971	396 944	3 353	10 782	22 835
Temperate South America	5 741	44 242	44 379	205	1 056	1 292
Tropical South America	47 296	136 943	294 654	2 369	7 591	16 038
Caribbean & Central America	8 414	24 786	57 911	779	2 135	5 505
Africa	237 883	481 096	514 671	28 700	58 700	79 875
North Africa	84 634	176 498	172 910	7 500	14 750	22 453
West Africa	53 600	181 324	204 982	9 000	22 000	29 478
East Africa	39 496	86 096	106 030	9 000	18 750	24 211
Southern Africa	60 153	37 178	30 749	3 200	3 200	3 733
Asia	143 876	412 937	507 113	86 900	142 500	148 780
West Asia	28 211	97 315	119 780	7 000	18 800	16 535
South Asia	56 422	87 727	82 194	57 000	74 000	80 958
Southeast Asia	19 747	55 441	63 959	9 500	30 500	33 987
East Asia	39 496	172 454	241 180	13 400	19 200	17 300
China	n.a.	n.a.	n.a.	n.a.	n.a.	n.a.
WORLD TOTAL	1 427 913	2 722 564	3 443 489	177 521	294 483	349 337

Sources: Boyce and Evenson (1975), Judd et al. (1983).

TABLE 3

Research and Extension Expenditures as a Percentage of the Value of Agricultural Product

Sub-region	Public sector agricultural research expenditures			Public sector agricultural extension expenditures		
	1959	1970	1980	1959	1970	1980
Northern Europe	0·55	1·05	1·60	0·65	0·85	0·84
Central Europe	0·39	1·20	1·54	0·29	0·42	0·45
Southern Europe	0·24	0·61	0·74	0·11	0·35	0·28
Eastern Europe	0·50	0·81	0·78	0·32	0·36	0·40
USSR	0·43	0·73	0·70	0·28	0·32	0·35
Oceania	0·99	2·24	2·83	0·42	0·76	0·98
North America	0·84	1·27	1·09	0·42	0·53	0·56
Temperate South America	0·39	0·64	0·70	0·07	0·50	0·43
Tropical South America	0·25	0·67	0·98	0·34	0·71	1·19
Caribbean & Central America	0·15	0·22	0·63	0·09	0·18	0·33
North Africa	0·31	0·62	0·59	1·27	2·21	1·71
West Africa	0·37	0·61	1·19	0·58	1·24	1·28
East Africa	0·19	0·53	0·81	0·67	0·88	1·16
Southern Africa	1·13	1·10	1·23	1·64	0·67	0·46
West Asia	0·18	0·37	0·47	0·25	0·57	0·51
South Asia	0·12	0·19	0·43	0·20	0·23	0·20
Southeast Asia	0·10	0·28	0·52	0·24	0·37	0·36
East Asia	0·69	2·01	2·44	0·19	0·67	0·85
China	0·09	0·68	0·56	n.a.	n.a.	n.a.
COUNTRY GROUP						
Low-Income Developing	0·15	0·27	0·50	0·30	0·43	0·44
Middle-Income Developing	0·29	0·57	0·81	0·60	1·01	0·92
Semi-Industrialized	0·29	0·54	0·73	0·29	0·51	0·59
Industrialized	0·68	1·37	1·50	0·38	0·57	0·62
Planned	0·33	0·73	0·66	—	—	—
Planned — excluding China	0·45	0·75	0·73	0·29	0·33	0·36

(Sources: Tables 1 and 2; World Bank, 1983a)

These manpower intensities should not be interpreted as though there were no differences in the quality of manpower between countries. There is little doubt that the general levels of training of both scientists and extension workers vary between countries, and are lower in the developing countries. However, there is little indication that these differences have changed as research and extension spending has increased. These data do not include spending of an 'extension type' associated with Rural Development Projects in developing countries; if

TABLE 4
Research and Extension Manpower Relative to the Value of Agricultural Product

Sub-region	SMYs per 10 million (constant 1980) dollars agricultural product			Extension workers per 10 million (constant 1980) dollars agricultural product		
	1959	1970	1980	1959	1970	1980
North Europe	1·05	2·01	3·14	2·76	2·56	2·61
Central Europe	0·80	1·21	1·56	2·19	2·77	2·73
Southern Europe	0·93	1·17	0·96	2·00	2·76	2·69
Eastern Europe	1·44	2·97	2·84	2·36	2·88	3·13
USSR	1·38	2·37	2·34	2·26	2·33	2·50
Oceania	1·91	2·64	2·43	2·26	2·17	2·11
North America	0·84	0·89	0·84	1·44	1·31	1·08
Temperate South America	0·46	1·15	1·32	0·26	1·19	1·26
Tropical South America	0·41	1·41	1·77	1·71	3·95	6·46
Caribbean & Central America	0·53	0·86	1·20	0·82	1·53	3·12
North Africa	0·91	1·44	4·24	18·83	28·45	22·23
West Africa	0·33	0·61	1·42	7·61	14·01	18·08
East Africa	0·32	0·77	1·76	16·28	22·41	26·64
Southern Africa	1·90	1·96	2·47	8·73	5·94	5·62
West Asia	0·33	0·84	0·88	4·39	7·25	6·54
South Asia	0·50	0·65	1·29	20·83	19·51	19·53
Southeast Asia	0·47	1·28	2·07	9·81	13·07	19·72
East Asia	3·80	5·29	5·72	6·57	7·05	6·13
China	0·22	1·66	1·49	n.a.	n.a.	n.a.
COUNTRY GROUP						
Low-Income Developing	0·43	0·67	1·40	18·14	18·61	20·43
Middle-Income Developing	0·69	1·31	2·40	8·89	14·68	15·98
Semi-Industrialized	0·70	1·21	1·36	2·80	4·95	5·21
Industrialized	1·24	1·71	1·85	2·37	2·31	2·12
Planned	1·02	2·27	2·13	—	—	—
Planned — excluding China	1·40	2·54	2·50	2·29	2·49	2·63

Sources: Boyce and Evenson (1975), Judd *et al.* (1983), World Bank (1983*a*).

such data were to be tabulated and included as extension spending, the magnitude of the differences in spending on extension relative to research in the developing countries would be even greater.

Table 5 provides further insight into the motivation for the high extension manpower intensities in developing countries. It shows expenditure: manpower ratios for research and extension. These ratios include salaries of scientists and extension workers and related costs, including laboratory costs and the costs of technicians. The ratio of research costs to extension costs is as much as 20 to 1 for the low-income developing countries, but only 3 to 1 or so for the industrialized countries. Some of this difference is a quality difference (extension workers have received quite advanced training in most industrialized countries, but many have had little training in low-income countries), and some is due to real cost differences. Many low-income countries do not have the capacity to train agricultural scientists, and must incur high costs to train research workers and to purchase scientific equipment.

These differences in cost, of course, reflect large differences in 'quality' of extension workers. In some low-income countries the available supply of extension workers with scientific training in agriculture is very limited. With programmes to expand extension systems rapidly, administrators have devised ways to build extension systems that can utilize the relatively untrained and unskilled field worker. Most extension development programmes have chosen to emphasize short-term training of both field and administrative staff as opposed to longer-term investment in skill development through graduate scientific education. For example, the Training and Visit system (T & V) is at least partly designed to deal with low skill levels.

Given the constraints facing managers of extension systems, it is economically rational that such systems should utilize low-skill manpower. Later, I show that governments do respond to low prices for extension workers by hiring more extension workers relative to research workers. It remains an open question, however, whether this is efficient.

STUDIES OF THE ECONOMIC CONSEQUENCES OF EXTENSION

A number of statistical studies examining the impact of extension programmes on farm productivity and income have now been made.

TABLE 5
Agricultural Research/Extension Expenditures per SMY/Extension Worker

Region/Sub-region	Research expenditures per SMY (,000 constant 1980 US $)			Extension expenditures per extension worker (,000 constant 1980 US $)		
	1959	1970	1980	1959	1970	1980
Western Europe	44	73	76	15	19	18
Northern Europe	52	52	51	24	33	32
Central Europe	49	98	99	13	15	16
Southern Europe	25	52	78	5	13	11
Eastern Europe & USSR	32	29	29	13	13	14
Eastern Europe	34	27	27	14	12	13
USSR	31	31	30	12	14	14
North America & Oceania	90	127	127	28	40	51
North America	100	142	130	29	41	52
Oceania	52	85	117	24	35	46
Latin America	56	44	54	18	19	18
Temperate South America	85	56	53	28	42	34
Tropical South America	61	48	56	20	18	18
Caribbean & Central America	28	26	52	11	12	11
Africa	62	65	53	8	8	6
North Africa	35	44	27	11	12	8
West Africa	108	97	83	6	8	7
East Africa	58	72	46	4	5	4
Southern Africa	59	56	50	19	12	8
Asia	23	38	39	2	3	3
West Asia	53	44	54	4	5	7
South Asia	22	28	34	1	1	1
Southeast Asia	20	22	25	2	2	2
East Asia	18	38	43	3	9	14
China	43	41	37	n.a.	n.a.	n.a.
COUNTRY GROUP						
Low-Income Developing	34	40	47	2	2	2
Middle-Income Developing	42	44	47	7	7	6
Semi-Industrialized	41	45	46	10	10	11
Industrialized	55	80	93	16	25	29
Planned	33	32	31	—	—	—
Planned excluding China	31	25	30	13	13	14

(Sources: Boyce and Evenson, 1975; Judd *et al.* 1983).

Most of these are reviewed in a recent book by Jamison and Lau (1982). I will summarize these studies briefly here. I will also discuss a more recent study (Evenson, 1985) based on international data. (Readers will also be interested in the study by Feder and Slade in this volume in this connection.)

The methodology for most of these studies is statistical. Multiple regression methods are used to estimate whether farmers with access to more extension services are more efficient and whether they have higher incomes. The statistical model utilized is either a 'production function' specification or a 'productivity decomposition' specification. The specification for the production function is usually 'linear in logarithm', and the specification for productivity decomposition is usually linear so that standard 'ordinary least squares' regression methods are used. The typical specification includes variables characterizing the public provision of extension services (usually expressed as extension time per farm); variables characterizing the technology associated with agricultural research (measured as a stock of cumulated past investments in research with a time-lag built into the variable usually expressed on a 'per unit of commodity output per geo-climate region' basis); and variables measuring the schooling of farmers.

Some specifications include 'interaction' variables. New variables are formed by the multiplication of the extension and research variables, and the extension and schooling variables. This allows a measure of the complementarity or substitutability of extension with research or schooling. For example, if the 'Extension × Research' variable has a positive sign, this means that the marginal impact of extension on production is larger if more research is being undertaken. We would then describe extension as complementing research productivity, and vice versa. If the 'Extension × Schooling' variable is negative, this means that the marginal impact of extension is lower the more schooling the farmers have received. This can be described as a substitution relationship.

Table 6 provides a brief summary of nine of the earliest studies of extension impact. These studies identified positive extension programme impacts on farmer productivity. All but the Brazil study claimed statistical significance for their estimates. The four studies by Huffman for US agriculture showed that when extension spending is treated as a public investment, the benefits from improved productivity are sufficiently high that returns to investment are well above the range of normal returns to investment. The studies on India, Brazil, Kenya

TABLE 6
Summary of Returns to Extension

Study	Country (data set year)	Type study	Conclusion
1. Patrick and Kehrberg (1973)	Brazil–Eastern (1968)	Production function	Extension, number of direct contacts of farmers with extension agents during the study year, had positive but generally not statistically significant effects on value added in farm production.
2. Evenson and Jha (1973)	India (1953–54 to 1970–71)	Productivity change	Extension, index of maturity of extension programme, contributes significantly to agricultural productivity change only through interaction with research programmes. Investment in extension programmes yields a 15–20 per cent social rate of return.
3. Huffman (1974)	USA — Corn Belt (1959–64)	Allocative efficiency-production	Extension (days, average for 1958 and 1960, allocated to crops by agents doing primarily agricultural work) and education are substitutes in inducing optimal nitrogen fertilizer usage on hybrid corn. The marginal value of extension time on this one decision is estimated at $4.48 per hour of extension agent time allocated to crops or a social rate of return of 1·3 per cent. Total social return from enhanced decision-making suggested to be in excess of 16 per cent.
4. Mohan and Evenson (1975)	India (1959–60 to 1970–71)	Productivity change	The Intensive Agricultural Districts Program (presence vs. absence) contributed to more rapid agricultural productivity change. The social rate of return realized on the investment was 15–20 per cent.

(continued)

TABLE 6—*contd.*

	Study	Country (data set year)	Type study	Conclusion
5.	Huffman (1976a)	USA — Ia, N.C., Okla. (1964)	Production function	Extension, agent days allocated 3 years earlier to crops and livestock activities by agents doing primarily agricultural work, contributes significantly to level of agricultural production. The marginal product of extension is $1 000–3 000 per day.
6.	Moock (1976, 1978)	Kenya–Vihiga, a western division (1971)	Production function	An index of crop related extension contact with male and female farm operators during the last year contributes significantly to corn (maize) yields. Extension and education are substitutes in corn production; extension interacts positively with the rate of nitrogen fertilizer application on male operated farms (1978).
7.	Huffman (1976b)	USA — Ia, N.C., Okla. (1964)	Production function	Same as for Huffman (1976a) except marginal product of extension $1 000–2 500 per day.
8.	Halim (1977)	Philippines — Laguna Province (1963–68–73)	Production function	An index of extension contact with farms, derived by weighting frequency of contact over previous 5 years, contributes positively and significantly to agricultural production. Marginal products imply a 'relatively high return of extension contact'.
9.	Huffman (1977)	USA — Corn Belt (1959–64)	Allocative efficiency	Same as Huffman (1974) except marginal value of extension time on this one decision is estimated at $600 per day of extension agent time allocated to crops or a social rate of return of 110 per cent.

and the Philippines, on the other hand, generally produced more modest returns to investment.

In another study of US agriculture, Evenson (1982) using state level data for 1949–71 also found a statistically significant extension impact on agricultural productivity. In that study a $1000 increment to extension spending was estimated to cause a $2173 increment to farm output within a 2 year period. This translates into a rate of return on investment of approximately 100 per cent. The study also showed that farmer schooling and research yielded high returns to investment as well.

The study also showed that the interaction between extension and farmers' schooling was negative. This means that extension programmes and farmers' schooling are substitutes. An increase in the level of farmers' schooling makes extension services less valuable. This finding is almost universal for all studies investigating this question.

The same study found a strong positive interaction with the research investments in US State Experiment Stations. Higher levels of research investment made extension more productive, and higher levels of extension investment made research more productive. This result, as noted below, is not universal. It appears to hold only in relatively advanced systems. In less developed countries it appears that extension does not complement research.

Jamison and Lau (1982) report several additional studies of extension impact. Studies from Japan, Korea, Malaysia and Thailand are reviewed in addition to those discussed in Table 6. These studies all found significant extension impacts, except for the Thailand study where no effect was found on farms using chemical fertilizer.

The complementarity relationship between research and extension found in the United States data is not consistently found in data for developing countries. This is shown in Table 7 from a recent study of 25 developing countries (Evenson, 1985) for the period 1962–80. This study sought to estimate the commodity-specific impact of the national research programme on the commodity, the national extension programme and the International Agricultural Research Center (IARC) programme on the commodity. The specification of variables was similar to that utilized in the national studies. Extension expenditures in this study are 'deflated' by the crop and geo-climate diversity of each of the countries.

Two items of interest for extension policy can be drawn from Table 7. The first is the interaction terms. The interaction between national research and extension is consistently negative in all commodities and

TABLE 7

Research and Extension Impacts on Crop Productivity, 1962–80

Research-extension coefficient	Maize, millets & sorghum			Cereal crops			Staple crops		
	Latin America	Africa	Asia	Latin America	Africa	Asia	Latin America	Africa	Asia
National Research	0·012 1**	0·039 3**	0·031 4**	0·014 6**	0·854 (3)	0·010 6**	−0·019**	0·073 3**	0·047 9**
National Extension	0·033 1**	−0·609 (4)	0·030 5**	0·015 8**	−0·153 (3)	0·038 9**	−0·493 (2)	0·939 (2)**	0·015 7*
Natl. Res. × Natl. Ext.	−0·117 (2)**	−0·939 (3)**	−0·172 (2)**	−0·364 (3)**	−0·228 (3)	−0·597 (3)**	0·318 (3)**	−0·101 (2)*	−0·457 (2)**
IARC Research	0·286 (5)	0·809 (5)	0·213 (6)	0·560 (5)**	0·319 (5)	0·171 (5)	0·237 (4)**	0·371 (5)	0·514 (5)
IARC Res. × Natl. Res.	−0·179 (6)	0·445 (6)	−0·103 (5)**	−0·193 (6)**	0·157 (7)	−0·644 (7)**	0·685 (6)*	−0·228 (5)	0·105 (5)
IARC Res. × Natl. Ext.	0·129 (5)**	0·178 (6)	0·349 (5)**	0·501 (7)	0·222 (6)**	0·755 (6)*	−0·737 (6)*	0·653 (6)	0·188 (5)
Productivity Elasticities									
National Research	0·034 4	0·050 5**	0·116 8**	0·143 5**	−0·006 0	0·113 5**	−0·030 2**	0·031 3**	0·129 2**
National Extension	0·170 8*	−0·012 9	0·165 8**	0·074 5**	0·012 8	0·192 1**	−0·024 3**	0·119 8**	0·068 5
IARC Research	0·031 7*	0·035 5**	0·041 6**	0·029 8**	0·054 3**	0·042 8**	0·041 2**	0·018 7	0·031 2*

Note: Numbers in parentheses are E(−n).
*'T' or comparable 'F' indicate significance at 5 to 10 per cent levels.
**'T' or comparable 'F' indicate significance at 5 per cent or lower levels.
Source: Evenson (1985).

regions. This shows that extension activities have a substitute relationship with research. More extension lowers the marginal product of research, and vice versa! The interaction coefficients between extension and IARC research, on the other hand, are generally positive. This suggests that there are very poor linkages between national research and extension programmes in most developing countries. The products of the research system are not translated effectively into productivity gains by extension services. The products of the IARCs, on the other hand, are enhanced by extension activity. The lack of complementarity between national research and extension is related to the organization of both research and extension. It is affected by the scientific and technical competence of both researchers and extension workers.

The lack of complementarity between research and extension does not indicate that these activities are unproductive. Indeed, the production elasticities reported in Table 7 (which indicate the percentage increase in production from a given percentage change in national research, extension, or IARC research) show that national research programmes have positive impacts on productivity for most commodities in most regions. The same is true for extension. The estimates show that IARC programmes have had strong positive impacts as well. The elasticities for extension generally imply moderate rates of return. The highest estimates (0·17 to 0·19) imply a rate of return of about 40 per cent if the country is spending 1 per cent of the value of agricultural product on extension. If 1·5 per cent of the value of the product is spent on extension as in some African countries (see Table 3) the highest elasticities imply only a 10 per cent rate of return. Thus, even where we seem to be observing highly productive extension services in developing countries, one cannot justify extension programme spending of much over 1 to 1·5 per cent of the value of the agricultural product. If the programmes are not highly productive they will not increase productivity sufficiently to pay for their costs.

STUDIES OF DETERMINANTS OF EXTENSION INVESTMENT

A number of studies in recent years have investigated the spending decision by governments on agricultural research and extension. Several studies of spending at the state level in the US (Peterson, 1969; Huffman and Miranowski, 1981; Rose-Ackerman and Evenson, 1985) have been undertaken. They show the following:

(1) Extension spending is more closely related to the state farm population than to its farm income (the reverse is the case for research).

(2) Political variables reflecting the power of farmer membership in legislatures and rural over-representation are associated with higher research and extension spending.

(3) States tend to 'free-ride' in both research and extension spending. This means that a state will respond to increased spending by a second state located in the same geo-climate region by spending less than it otherwise would.

(4) Increased federal funding to a state causes the state to spend more on both research and extension even though it is not required to do so by the terms of the funding.

Of more relevance to this paper, however, is a recent international investment study (Evenson, 1985) based on the 25 country data set discussed in the previous section. The study sought to determine factors influencing national research and extension spending. Of special interest was the impact that IARC programmes had on national spending. Were countries responding to IARC programmes by 'free-riding', i.e., spending less than they otherwise would? Were they 'free-riding' on the research of other countries? Does aid from the World Bank or other agencies actually stimulate an increase in national research and extension spending?

Table 8 summarizes the main results of the analysis. Here we present estimated impacts of several factors on national research expenditures for field crops, for livestock and horticultural crops, as well as for extension.

The estimates show that both research and extension spending elasticities for commodity production are around 0·6. This means that as countries become larger, in terms of the value of their product, research and extension spending increase less than proportionately. A 10 per cent increase in production is associated with a 6 per cent increase in spending. This reflects 'fixed' costs and scale economies in running programmes.

When additional commodities are exported, this stimulates more research and extension spending by one and a half to twice as much, compare lines 1 and 2 in Table 8. When more commodities are imported, research spending is increased, but extension spending declines (line 3). Countries respond to increased research by geo-climate neighbours by spending more on research and less on extension (line 4). When relative agricultural prices are more favourable to farmers (i.e., when the ratio of

TABLE 8

Calculated Impacts on National Research and Extension Investment (Calculated Impacts on Spending in Millions of 1980 Dollars)

Policy variable	Research spending (million dollars)		Extension spending (million dollars)
	Field crops	Livestock and horticultural crops	
(1) 1 million dollars added to commodity production (elasticity)	0·001 64 (0·551)	0·003 96 (0·584)	0·006 24 (0·592)
(2) 1 million dollars added to commodity exports	0·000 634	0·002 277	0·006 95
(3) 1 million dollars added to commodity imports	0·000 472	0·012 53	−0·000 937
(4) 1 added SMY by geo-climate neighbour	0·030 5	0·019 01	−0·179 2
(5) Ten per cent decline in urea-rice price	−0·132	0·342 5	1·414
(6) Ten per cent decline in research costs per SMY	−0·171 6 to	−0·423 to	1·452 to
(cost ± std deviation)	0·046 2	−0·638	1·878
(7) 1 million dollars added to IARC research stock			
(a) first year	0·229	1·084	0·105
(b) after 10 years	2·290	10·840	1·050
(8) 1 million dollars general aid research	1·194	−0·858	0·047
(9) World Bank aid (to research or extension)	0·285	−0·063	1·468

the price paid for urea fertilizer to the price received for rice falls), more is spent on extension (line 5).

When the cost of researchers falls or the cost of extension workers rises, countries respond by hiring more researchers and fewer extension workers. A 10 per cent rise in the cost of fielding an extension worker induces a 15 per cent decrease in the number of extension workers hired and a 5 to 10 per cent increase in the number of researchers (line 6). IARC spending stimulates national spending for both research and extension. The cumulated impact after 10 years is large (line 7). General aid to research stimulates countries to 'shift' research from livestock and horticultural crops to field crops, and has no appreciable impact on extension. World Bank aid to research has a similar effect. General aid to research is mostly 'displaced', i.e., one million dollars causes net added spending of less than 350 thousand dollars (the sum of the first two coefficients in line 8). World Bank aid to extension has a powerful leverage impact. One million dollars in aid (or loans) causes an expansion in domestic spending of more than a million dollars (line 9). This is presumably related to the Bank's handling of T & V extension aid.

IMPLICATIONS FOR EXTENSION PROGRAMME DESIGN IN DEVELOPING COUNTRIES

The economic studies reviewed here do not provide full guidelines for efficient extension system design. They do indicate the following:

(1) Extension programmes in developed countries are generally substitutes for farmers' schooling and complements to research programmes. Public sector expenditures as a fraction of the value of agricultural product are in the 0·6 to 0·8 range. These programmes appear to be highly productive.

(2) Extension programmes in developing countries are highly variable. Studies show that extension and schooling are substitutes but they also show that extension programmes do not complement research programmes. Evidence on rates of return are variable. It appears, however, that even the more productive developing country programmes cannot justify investment levels of much more than one per cent of the value of agricultural product.

Reference back to the discussion centring on Fig. 1 will be helpful in interpreting some of the evidence reviewed here. Many developed countries have managed to develop a complete portfolio of productivity enhancement institutions. Pre-technology science research complements applied research and applied research complements extension in these systems. The entire system avoids the 'exhaustion' that sets in when upstream components complement downstream components.

For many developing countries we observe that research programmes are poorly developed and that they may provide little recharge to extension systems. We also observe, however, that many extension programmes in developing countries lack the levels of competence and the institutional links with research organizations to effectively achieve complementarity. When extension staff do not have a minimal level of scientific and technical competence they cannot effectively translate research findings into sound advice to farmers. When extension organizations are not closely linked institutionally with research organizations they will not function effectively.

It should be acknowledged, of course, that extension systems can be productive even when they do not complement research programmes. This is particularly true in the early stages of development where simple adult education is highly valuable. Extension services have always had a strong and usually productive role in providing basic educational services that do not necessarily depend on recent or new research results. This role, however, is subject to exhaustion after some time, and if extension services do not develop a capacity to convey technological information from upstream they will be of marginal value after a period.

Many extension programmes in developing countries were built by hiring large numbers of poorly trained and, in many cases, untrained staff. It seems likely that many extension agents lack the minimal qualifications to be effective educators or technology advisers. Organizational mechanisms such as the T & V system can enhance the effectiveness of such systems but cannot fully overcome the problems associated with low competence.

To date, we have conducted relatively few economic studies of extension programme impacts. We have evidence that extension programmes can be an excellent social investment if properly organized and if other complementary investments in research and schooling are made. We have evidence that some investments in extension have not yielded high returns. Continued study is called for to improve our understanding of the types of effective extension investment that can be made in developing countries today.

POLICY, AND THE PLANNING OF EXTENSION

5

Extension Policies, Policy Types, Policy Formulation and Goals

ANNE W. VAN DEN BAN

Department of Extension Education, Agricultural University, Wageningen, The Netherlands

Agricultural extension is one of the policy instruments which a government can use to stimulate agricultural development. The question is: for which purposes is this the most suitable policy instrument, and how can it be combined with other policy instruments, such as agricultural research, vocational agricultural education, the provision of credit and supplies, marketing, the organization and mobilization of farmers, and infrastructural works (such as roads or irrigation)?

GOALS

Most extension organizations try to achieve several goals. However, the emphasis on the various goals differs from country to country. An indication of this variation are the words used for extension in different languages.

The goals of extension organizations include:

— the *transfer of knowledge* from agricultural research to the farmers. This is stressed in the French word *vulgarisation* and in the term 'extension';
— *advising* farmers on the decisions they have to make, sometimes by recommending a certain decision to be taken, sometimes by helping them to acquire sufficient insight into the consequences of the alternatives from among which they can choose in order that they can make their own decision. In the UK, Germany and the Scandinavian countries one speaks of 'advisory work'; in the

Netherlands and in Indonesia we speak of *voorlichting* and *penyuluhan*, which indicate keeping a light in front of the farmer;

— *education,* helping farmers to make a decision in such a way that they are able to make similar decisions themselves in the future; This is implied in the US term 'extension education';

— *enabling* farmers to find their own way by helping them to clarify their goals and the possibilities which they have, together with other farmers, to realize these goals; The French word *animation* and the Spanish word *capitación* point in this direction;

— *stimulating* desirable agricultural *developments,* as is implied by the Austrian word *Förderung* (furthering) and the Korean expression 'rural guidance'.

Which of these goals receives most emphasis depends partly on the views of the farmers on which the extension policy is based. In industrial psychology we have seen a change from McGregor's theory X towards theory Y and to some degree back again; theory X implies that workers will only work if they are forced to do so, whereas theory Y maintains that people are mainly motivated by their desire to do their job well. How much confidence do we have that farmers are able and willing to become good farmers? Farmers might react to a lack of confidence which extension workers have in them by a lack of confidence in the extension workers.

The choice of the goals of an extension service also depends on the reasons why one believes that the desired kind of modern agriculture has not yet been achieved. To what extent is this due to a lack of knowledge among farmers of the results of agricultural research? To what extent is it because agricultural research has not yet produced findings which are applicable in the situation of most farmers? To what extent are other elements of the agricultural development mix (inputs, credit, transport, markets, etc.) lacking in the villages? If they are lacking, can they be provided most effectively by the government, by farmers' organizations, or by private business?

If farmers' organizations and co-operatives should play a role in bringing part of this development mix to the farmers, then farmers should learn how to organize themselves effectively. These farmers' organizations could also play a useful role in reducing the urban bias which exists in development policies in many countries. Is it the task of the extension service to help farmers to learn how to organize themselves effectively, or should this be a function of some other governmental or non-governmental organizations? This is a politically sensitive question, but in my opinion it is also an important question. In many industrialized countries we see that farmers' organizations, run by the farmers themselves, are playing a very important role in the economic and political life of the country. I am convinced that without these organizations the present level of agricultural development could never have been reached.

STIMULATING DESIRABLE DEVELOPMENTS

My impression is that most directors of agricultural extension in developing countries see the stimulation of desirable developments as the main goal of their extension work. To use extension as a policy instrument to stimulate these developments raises several questions:

(1) Which developments are desirable? In industrialized countries, only about 4 per cent of the population is now employed in agriculture, and this small minority produces enough food for the whole population. Of course, it requires the use of considerable capital and chemical inputs. Should the developing countries move in the same direction? If so, how can they provide employment for those people who are pushed out of agriculture through this development?

(2) In many countries there are large differences among the farmers in their access to resources and in education. Are the same developments desirable for all farmers? If not, which group should the extension service try to help most? This last question is often not answered by the policy makers, but it is by those farmers who ask the extension workers to help them.

(3) Farmers are free to follow or not to follow the advice offered to them by their extension workers. They will only follow advice if they are convinced that this is in their interest. At the same time the government will only pay for the extension service from the taxpayers' money if this service contributes to changes which are in the national interest. In which circumstances does this national interest coincide with the interest of the farmers? Only in those circumstances does extension education seem to be the correct policy instrument to realize the desired changes. It can,

for instance, be in the interest of the farmers as well as in the national interest to increase crop yields by the use of better seeds. If it is in the national interest to increase the production of export crops, but at current price levels not in the interest of the farmers, this goal cannot be realized by extension education. In several countries the extension services have been expected by the politicians to realize changes which were not in the interest of the farmers. Usually this has resulted not only in a failure, but also in a loss of confidence by farmers in their extension workers.

(4) To what extent are the policy makers an important target group for the extension service? In order to design realistic policies, policy makers need a good understanding of the situation of the farmers and of the reasons why they do what they do. Good extension workers have much of this information, but not all policy makers are interested in learning from them.

SHOULD THE TAXPAYER PAY FOR AGRICULTURAL EXTENSION?

In many countries, governments are paying for large extension services for agriculture, but only for small extension services for small businesses. In industrialized countries we see that commercial companies, farmers' organizations and private consultants are responsible for a large proportion of the extension advice to farmers. It is therefore reasonable to wonder whether tax revenue should be used to finance an agricultural extension service.

One reason for a government extension service is the international competition which exists in agriculture, which is not present for many kinds of small businesses. Suppose that the extension service is increasing the efficiency of production, then a good extension service enables farmers to compete better with farmers in other countries. This is probably one of the reasons why the decrease in the agricultural labour force in recent years is considerably less in the Netherlands than in other EEC countries or in the US. Agricultural extension and other support services for agricultural development have enabled Dutch farmers to compete well with farmers in other countries.

Another reason why a government pays for agricultural extension is that the main effect of increased efficiency in agricultural production is not increased farm incomes, but rather decreased consumer prices for agricultural products and less chance of food shortages or a famine. In this situation it is fair that consumers should also pay for the extension service which helps to increase the efficiency of agricultural production. The story becomes different, however, if the agricultural policy prevents this decrease in consumer food prices.

In industrialized countries the trend is for governments to supplement the extension work done commercially. This supplementation is especially needed where the interests of the commercial organizations and the interests of the farmers do not coincide. It can, for example, enable farmers to check whether they have received correct advice from a commercial company. Also, the educational role of increasing the managerial abilities of the farmers remains an important task for the government.

In many less industrialized countries, poverty alleviation is nowadays an important goal of the agricultural development policy. Commercial extension organizations are not interested in helping poor subsistence farmers. This can only be done by the government, but it requires that extension workers are rewarded for working with these farmers, who are often hard to reach. An extension service can be a means of increasing the capabilities of small farmers to compete with plantations and other large, modern farms.

STATUS OF AGRICULTURAL EXTENSION

In many countries, agricultural extension work has a relatively low status. This makes it difficult to attract really capable staff. Agricultural faculties do not attract many of the best students, while agricultural graduates prefer to work in research, in the ministry, or with private companies rather than in the rural areas as part of the extension service.

The low status also means that the agricultural extension service finds itself in a weak position in the fight between different government agencies for an adequate budget. However, the status of the agricultural extension service is not low everywhere. In some countries the linkage of the farmers with agricultural research through the extension service is seen as an example of effective research utilization which should be followed in other branches of the economy in order to stimulate economic growth.

The status of an extension service probably depends on:

(1) the extent to which it is seen to contribute to important societal goals. In many African countries the contribution to an increased food production is now considered to be more important than it was ten years ago;

(2) the extent to which it fulfils this role in a professional way by using capabilities which are not available outside the extension service. This can be professionalism in agriculture, which might stimulate the service to focus on the modern rather than on the subsistence farmer. It can also be professionalism in communication strategies by which it is possible to change the behaviour of traditional farmers;

(3) the extent to which it is well regarded by people with a high status in the society, for example, by large-scale farmers.

Research by Evenson has shown that investments in agricultural research and extension often produce a higher rate of return than investments in infrastructural works. The amounts invested in these works are, however, much larger than the amounts invested in extension and research. I am convinced that a small decrease in infrastructural investments works (e.g. in irrigation works), if used for what would be a large increase in investments in extension and research, could increase the rate of agricultural development. However, this would be the case only if agricultural extension and research are organized in an efficient way. That is not the case everywhere. Recently, we have seen that many countries have considerably increased their investments in agricultural extension by adopting the Training & Visit system. Through this system they have also tried to increase the efficiency of their agricultural extension service. Investments in infrastructural works, however, remain politically more attractive because they are more visible.

THE COMBINATION OF EXTENSION WITH OTHER POLICY INSTRUMENTS

It is clear that agricultural extension should be combined with other policy instruments to achieve agricultural development. Often, extension workers are also responsible for some of these other policy instruments, such as input supply and credit. This is now usually discouraged, because the result is that extension education is invariably neglected. It can be delayed much easier than, for example, the supply of inputs. Moreover, extension workers can easily lose the confidence of farmers if the input supply is not well organized or if they have to force farmers to repay their debts.

However, it is relevant to ask who else can take the responsibility for input supply, credit and marketing in the early stages of agricultural development? The demand for these services is so small at this stage that it is not sufficiently attractive to expect commercial companies to provide them. Can the extension organization teach groups of farmers to organize these services for themselves? If so, how can some powerful farmers be prevented from using these organizations to exploit their colleagues? Or, can we afford to organize separate government agencies to provide these services? In either case, can we coordinate agricultural extension with the other policy instruments to stimulate agricultural development? In many countries, serious bottlenecks to agricultural developments have arisen due to issues such as these being neglected or unresolved.

In discussing the relationships between policy and goals in extension work, this paper has been able to offer few answers to the questions raised. But, it is important to bring the questions into open debate as a basis for considering the validity and value of extension activities.

ACKNOWLEDGEMENTS

In preparing this paper I have profited from comments by Mr B. Huizinga and Mr M. J. Rolls on an earlier draft.

6

Agricultural Policy and Agricultural Extension: The African Experience

STEVE WIGGINS

Department of Agricultural Economics and Management, University of Reading, UK

This paper explores the links between agricultural policy and extension, and how they affect the objectives and management of extension services. Most of the comments made are drawn from African experiences, and may exaggerate the difficulties faced in other areas. Given the critical state of African agriculture, and the high proportion of Africans who depend on agriculture for their livelihood it seems fair to put Africa's situation before those of other regions. There is a further bias in this paper, and that is towards extension as carried out by Ministries of Agriculture. The relation of policy to agricultural extension is best described by looking initially at the sharp end of the extension system, the extension agents, and how policy affects them. The paper will then highlights those policy points which affect extension more or less directly, stressing the importance of context, and drawing out the implications for improved policy-making for agricultural development. In the process it should become apparent that the policy–extension linkage itself is not the critical point of intervention: rather, improvement lies in the more general domain of policy-making and management.

In Sub-Saharan Africa extension agents are all too often sad figures, abandoned in the bush with little or no support, infrequently supervised, with no messages worth passing on to farmers, and with few incentives to get on with the work. Farmers do not appreciate the agents' work, and only make use of them when they can gain access to some input — especially if subsidized — such as seed, fertilizer, chemicals, tractor hire, or farm credit. Consequently agents are demoralized, with little enthusiasm for their jobs.

Extension agents suffer from a triple bind. First, the policy they are meant to implement comes in an unregulated gush from Ministry of Agriculture (MOA) Headquarters via the District Agricultural Officer with scant regard to local specifics, co-ordination with past or current policies, the feasibility of achieving any target set, and the existing work-load of the agents. Usually there is no consultation with field agents or their supervisors on how policy should be implemented. Agents remain passive in the face of a sludge of half-baked instructions, often aware that they make little sense but powerless to do other than try and implement regardless. Recommendations from research stations typically leave much to be desired, being all too often formulated with little regard for smallholders or for the specifics of the extension agent's area. Agents find themselves asked to transmit recommendations that may at best be worthy but in need of adaptation to local conditions, and at worst are useless.

Second, agents working with subsistence-oriented smallholders find themselves having to advise on farming where the technical dimensions are especially difficult and locale-specific — in crop varieties, soil types, pests and diseases, and in farming systems.* To this farming environ-ment where so much depends on the local conditions, and where the local economy and society may be so adapted to them that changes in one part of the system may have unwelcome side-effects in another part, extension agents bring limited technical skills; at most they will be diplomates with two years' agricultural training, probably from a training oriented to high-input commercial farming associated with large farms and a modified environment. It is unlikely that agents have been taught much about smallholder farm systems, about local crop varieties, about low input farming, about dryland agriculture in semi-arid areas, or about unfashionable crops (e.g. millet, sorghum) and unfashionable animals (e.g. pastoral cattle, sheep and goats).

Third, agents operate at the far end of a crumbling bureaucracy. Typically, an MOA acts as a bureaucracy with a penchant for central-ization of decision-making, strict adherence to hierarchy, and authori-tarian relations between the hierarchical layers. It is concerned with the adherence to impersonal norms, with the avoidance of mistakes, and

---

*In contrast, large-scale commercial agriculture with its more limited range of enterprises, its comprehensive transformation of the environment (fertilizers to improve soils, pesticides to render crops sterile, machinery to level land, etc.) requires a narrower range of technical knowledge. Moreover that knowledge may be applicable over large areas for any given crop, for example irrigated rice, cocoa trees.

with close control of the actions of its staff. Promotion depends on paper qualifications and seniority: little attention is paid to performance, other than in the negative sense that staff should not have infringed some regulation. For extension agents, career structures exist barely if at all. Unfortunately, bureaucracy is a poor organizational form for farm extension work, where the agents in the field need to be motivated and prepared to take initiatives. If this were not bad enough, in many African countries the bureaucracy, including the MOA, is crumbling for lack of funds. Hence, staff allowances, and sometimes also salaries, are paid late and only after at least one visit to the district headquarters. After paying staff, the MOA has little to devote to supporting materials and facilities. So extension agents can do little that requires funds or other material support.

This bleak scenario* — perhaps a caricature of the worst that happens, but nevertheless true in large part for far too many field agents — results from the interplay of many factors. As far as policy is concerned, the following points can be made:

(i) Much agricultural policy can be described as uncoordinated, *ad hoc,* and arbitrary: in part, it is a reflection of political processes, in lesser part, the result of technical inability to formulate better policy. Moreover, policy objectives are frequently expressed with little regard to feasibility, manifestations of the 'we must run while others walk' school of politics (Hyden, 1983: Chap. 3). Objectives may not be feasible because the farming community could not achieve the targets set for

*There are exceptions, some notable, to this picture, for example:
(a) where agents have been employed to extend a single crop, so that they have been able to develop considerable knowledge and expertise;
(b) where the organization employing the agents has limited and specific objectives, very often the case with parastatals and private companies concerned with single crops;
(c) where explicit attention has been paid to setting up staff systems that encourage and reward agents for high performance; and
(d) where MOA agents have been used to carry out campaigns with a simple technical message.
In Kenya, good examples come from the Kenya Tea Development Authority (KTDA) and from the British-American Tobacco Company (BATCo) where conditions (a) to (c) prevail. In both cases their extension workers have been highly successful in encouraging smallholders to grow tea and tobacco, respectively.

lack of inputs, incentives, etc., or because the public agencies could not carry out the activities necessary. Moreover, policies on crops to be grown and husbandry practices are frequently formulated nationally, with no regard to the peculiarities of different areas. There may be no mechanism to rectify such over-simplifications, especially in a highly centralized administrative system.

(ii) Implementing policy may be made more difficult still by lack of correspondence between objectives and the allocation of resources. Some discrepancies result from deficiencies in the budget system (Leonard *et al.*, 1983), but more important is the fact that policy-making continues in the resource allocation process regardless of the policy already sanctioned by formal authority. In the murky world of resource allocation the covert objectives of agricultural policy emerge. For extension, a good example is that of providing employment for school leavers. In many countries, extension agents are the second largest group of public employees after schoolteachers.

(iii) Policy for research has low priority in agricultural policy making. Beyond general statements about the need for research, and research oriented to the needs of smallholders, little attention is paid to specific policy to achieve a flow of appropriate recommendations from research stations. To date, national research stations in Africa have tended to develop ideas with too little attention to smallholder labour supplies, to the riskiness of the innovations, to the likely availability of inputs, or to the presence of markets and to the economic attractiveness of recommendations. To an extent, Farming Systems Research (FSR) is helping improve the direction of research, but there is still one aspect which gets forgotten, and that is the link between research and extension. At worst, the extension service makes no use of research results,* and there is no feedback to the research workers of farmer reactions to recommendations.

(iv) Even less attention is paid by policy makers to extension itself, with the notable and recent exception of the debate about the

---

*In Tanzania, for example, the National Maize Project demonstrated that recommendations should be disaggregated by six ecozones, whilst the extension service persisted with a uniform national message which was patently inappropriate for some of the ecozones (World Bank, 1983*b*).

Training and Visit system. The existence of the extension service is not questioned, and the role of field agents is ill-defined. Agents are liable to be engaged in performing any task which fulfils Ministerial policy at village level, be it supplying inputs and credit, transferring technology, feeding back information to research workers, mobilizing local communities for group action to solve community-wide problems, or dealing with specific farmer problems and referring them to specialists. Agents become the Ministry in their villages: they must therefore do locally what the Ministry tries to do nationally. Because policy objectives tend to outstrip the resources available to achieve them, this leads to overload on the agents. Moreover, it also leads to them trying to do jobs for which they have neither the training nor the experience. The resultant pressure of being expected to do more than they are able both quantitatively and qualitatively demoralizes the extension staff.

(v) Policy for the extension service itself is dominated by targets of agent:farmer ratios, leavened by the secondary objective of raising the level of basic training the agents possess. Little thought lies behind such ratios: they have generally been taken from the ratios applying in North America or Europe. They are not matched to a model of how the extension worker is expected to affect the local community, which in turn would be difficult to conceive when the role of the agent has not been defined beyond statements about agricultural development so general as to be devoid of operational content.

These problems of policy as it applies to extension have been couched in general terms because greater specificity depends so much on context. There are three aspects to the context of agricultural policy. First, there are macroeconomic considerations of the role of agriculture in the economy and society. Similarly, the nature of economy — the degree of institutional development, the amount of rural infrastructure in place, the vigour of the private sector, etc. — plays a role. Second, for public sector activities like extension the capacity of the MOA (or parastatals), and its distinctive competence both technically and organizationally, including questions about organizational culture, constitute another aspect of the policy environment. Last, but not least, the type of local agriculture, the farm systems adopted, and variations from area to area are important considerations. The specificity of agricultural development puts a premium on policy-making and implementation which is disaggregated spatially. Hence, to be concrete about policy and extension requires an appreciation of the national economy, the capacity of the MOA, and the relevant farm system. Nevertheless, a number of points which policy makers and managers in agriculture would do well to consider apply in many circumstances:

(a) There is a need to define the role of the extension service in terms of the needs of farmers, the possible results of extension work, and the number and quality of staff, to mention only the main variables to be taken into account. Such a role may well vary from area to area, and over time in the same area.

For example, in a remote area of bush fallow farming, with few agents, the role of the extension service might be to provide supplies of seed and fertilizer unavailable from the small private sector. This would make slight demands on the knowledge of the staff and would use the extension service to relieve a critical bottleneck. On the other hand, in an area where much of the farming is medium to high input commercial agriculture, supported by a well developed infrastructure of input supply and marketing, the extension agent's role might be to educate farmers on the latest research results and their application to their farms, and to refer problems of pests and diseases back to specialists.

(b) More specifically, when defining extension's role, the link between extension and research needs to be considered in detail. Mechanisms like monthly meetings between extension administrators and research administrators, like as not held late on a Friday afternoon as everyone's lowest priority, are not enough. Unless the mechanisms are there, then there is no reason to expect any useful link between research and extension workers.

Possible mechanisms include: bringing the extension service in the immediate vicinity of research stations under station management; attaching research workers to extension agents for two days every month; having research stations conduct regular three day seminars for agents on recent recommendations, farmer reactions to them and on current research; and, creating senior posts for extension-research coordinators, one per major research station, whose job it would be to service such links and to create others, both formal and informal.

(c) The management of extension needs a great deal of attention. To induce staff to work with a will when they are isolated, scattered, for most of the time invisible to supervisors, in difficult areas and facing difficult professional problems, is hard. To get motivation and to ensure that agents are doing an effective job is even harder. The blind application of civil service rules designed to control the lives of filing clerks in Ministry Headquarters is not good enough. Better management of agents involves using alternative procedures for staff management — critically for promotion, job advancement, and supervision — and probably also demands more middle-level management.

(d) Policy needs to be better coordinated between aspects like pricing, input supply, marketing, and research and extension. It is not necessary to integrate actions in these different areas — indeed, to do so often means to create an unmanageable monster — but it is necessary to relate systematically different aspects of agricultural policy to one another. All too often what is seen in the field as a lack of coordination in implementation is a result of uncoordinated policy.

As a concluding comment, the recent introduction of the Training and Visit (T & V) system to Africa is to be welcomed, for a number of reasons. First, it brings extension to the attention of decision makers and opens a long-delayed debate on the use of the extension staff. Second, it provides a framework for better supervision of staff, an essential step on the road to better management. Third, it aims to strengthen the weak link between research and extension. T & V is no panacea, and already the critics (see Moore (1984) for a good example) have fired some damaging salvoes, but the principles of T & V are commendable. If these can be applied to specific contexts with appropriate modifications, then T & V will prove invaluable. If it fails in Africa, then I fear that it will merely have demonstrated just how intractable are the problems of agricultural development on that continent.

ACKNOWLEDGEMENTS

I am indebted to the ideas of Jon Moris and John Howell (see References) for some of their work germane to this topic, and to the comments of Deryke Belshaw and John Howell.

7

Flexibility versus Targets in Setting Extension Objectives

NIELS RÖLING

Department of Extension Education, Agricultural University, Wageningen, The Netherlands

INTERFACE AND BUREAUCRACY

Flexibility versus targets in setting extension objectives refers to an interface problem between the intervening extension system and its target system. Target systems do not always react in the manner expected. In addition to the question 'How do I get them to where I want them?', an extension worker must often ask the question: 'Why don't they do what I want them to do?' Usually the answer given is that farmers are backward, uncooperative, ignorant, traditional, or fatalistic. Progress out of this no-win situation becomes possible only if extension workers realise that farmers are like customers who do not necessarily have to buy what extension offers them. Then the answer to 'why don't they?' becomes: 'We obviously did something wrong, let's find out how we can improve our performance'. It is from this point of view that flexibility becomes important. Extension must have room for strategic manoeuvre; it must not paint itself into a corner by setting targets.

Trying to change an intentional target system usually requires flexibility. But flexibility has a cost. It is difficult for an extension bureaucracy to be flexible. It is difficult to plan for flexibility, difficult to supervise staff when objectives are flexible, and difficult to evaluate impact without clearly defining targets.

Therefore, is it possible for extension to be as flexible as the interface situation requires and yet function efficiently as a bureaucracy?

CONSEQUENCES OF SETTING TARGETS

Specifying targets in situations which are insufficiently known and which cannot be adequately controlled can have negative consequences. For example, setting targets in terms of tonnes per acre can lead to an extension focus on farmers who hold large acreages, while targets set in terms of numbers of farmers who adopt ensures wider coverage.

The targets which one sets determine the feedback one gets. Where impossible or unrealistic targets are set, feedback is deliberately misinformed. The top becomes blind to what happens in the system. Such targets corrupt the system. Realistic target-setting requires much attention to feedback, intelligence gathering and pre-testing (for instance, to know whether diffusion processes are taking off). Since extension does not live by its results, and is not funded or rewarded according to its performance, but often more by mistaken beliefs about what it should and can do, few incentives for realistic target-setting are provided. In such conditions, flexibility would seem to be called for.

However, the need of the apparatus for financial procedures which specify outlays twelve or more months in advance, for staff allocation with long lead times, for administrative control so as to allow for accountability for public funds, for responsiveness to the demands of the political system, and the power of the apparatus to arrange things according to bureaucratic convenience, all mitigate against realistic target-setting or flexibility.

THE NATURE OF THE EXTENSION INSTRUMENT

Extension is a communication intervention. It is first of all the profession of people who are paid to intervene, i.e., to use resources strategically to manipulate seemingly causal factors in a social process to change that process in a direction deemed desirable by the intervening party. But extension is limited to communication for leverage. Its objective is to change voluntary behaviour, i.e., to make people voluntarily do what one wants them to do. Much of the behaviour which is considered desirable by policy is of this nature, and communication (persuasion, information, instruction, feedback) is used to induce it. It is obvious that an intervention which can only be effective by inducing voluntary behaviour is somewhat of a paradox. Such an intervention can only be effective within certain rather narrow limits, or given certain conditions. Extension objectives, therefore, must take into account what can be feasibly achieved.

Understanding extension as an instrument of change allows specificity in setting objectives, while lack of understanding can be offset somewhat by flexibility. Specificity usually requires careful research into, or experience with, target conditions. Flexibility is safer in unknown conditions. Unfortunately, there usually is a strong correlation between specificity of targets and lack of experience and/or research data of the target situation.

THE PERSON 'IN-THE-MIDDLE'

In extension organizations, top administrators are usually close to politicians and policy makers. From their perspective, extension is a policy instrument to which public resources are allocated to achieve certain collective purposes. They set extension targets from this perspective.

The lowest ranks are occupied by people who have direct contact with farmers, who are often very similar to (progressive) farmers, and who know that they can only be effective if they induce voluntary behaviour, i.e., if they serve farmers' needs. They are 'in-the-middle', between the farmer and the top. As a result, there usually is a gap between the top and field workers which must be spanned by middle-level managers and organizational compromise. In situations where farming develops effectively, field workers succeed in thwarting the decrees from the top. When agricultural development is a failure (usually when farmers have no political power), the apparatus does not allow field workers to serve farmers and does not succeed in eliciting the voluntary behaviours required.

Political expediency and collective needs can easily lead to unrealistic targets and, consequently, to counter-productive effects, if enforced. Flexible objectives, combined with a strong claim-making capacity on extension by farmers, can lead to rapid agricultural development (including loss of employment opportunity in agriculture). Such an approach is, however, difficult to combine with pressure for accountability of extension to the political top.

THE LEVEL OF AGGREGATION

Extension is supported out of public means because it serves collective functions. It is usually called upon to provide cheap food for non-farmers, to provide foreign exchange through export crops, to provide cheap raw materials for industrial development, or to provide a surplus which can be creamed off for tax purposes or for supporting employment in agribusiness. Increasing farmer incomes is usually a necessary condition or by-product (in the case of earlier adopters). In other words, extension objectives are usually formulated for the collective level, not for the individual level. The interests of the nation, however, do not always coincide with farmer interests and are sometimes clearly opposed to them. Specifying extension objectives is often a question of knowing how much farming 'will bear'.

But extension effectiveness depends on it being able to provide farmers with assistance in reaching their goals. At the level of the farm, extension objectives formulated at the national level, therefore, often make no sense. In this situation the specification of targets means calling for trouble.

THE AGRICULTURAL DEVELOPMENT MIX

Extension is usually part of a mix of instruments. It is a weak instrument standing on itself, but becomes powerful especially when combined with price incentives, input distribution, credit, seed multiplication and so forth. The concept of a mix refers to the fact that the different instruments in the mix are not substitutes for each other but play complementary roles. The effectiveness of the mix is determined by the weakest or the absent instrument. A carefully assembled and tested mix provides much more intervention power and relies much less on voluntary behaviour change than extension alone. Targeting becomes more feasible with a mix of instruments.

TYPES OF EXTENSION OBJECTIVES

Objectives have been sequenced in many different ways. A well-known one is used in the logical framework: ultimate objectives — intervention objectives — performance objectives or activities — effort objectives or means. It is easier to specify targets at the level of performance objectives than at the level of intervention objectives, let alone at the level of ultimate objectives, to the achievement of which extension only makes a contribution. Effort objectives are usually very clearly specified and set limits to what is possible to begin with.

The sequence of objectives given above implies consistency: the activities performed must lead to achievement of the intervention objectives, and reaching these must contribute to achieving ultimate objectives. The effort made must be able to support the performance required. This internal consistency in the sequence means that extension programming must use adequate models of causal relations between activities, intervention outcomes and ultimate objectives. The better such causal models, the more it becomes possible to specify targets also at the higher levels of objectives in the sequence.

Flexibility in the performance, intervention, or ultimate objectives often requires flexibility in the effort objectives also. Financial and manpower resources, equipment and other means cannot be all determined beforehand if flexibility in dealing with a reactive environment is to be achieved. This is the greatest bottleneck to realizing any but fairly insignificant flexibility.

BLUEPRINT VERSUS PROCESS PLANNING

Setting clearly specified targets based on well-specified means implies that the 'present is known, the future knowable and that one can control events sufficiently to achieve a knowable future'. These assumptions underlie blueprint planning, so called because it is used in engineering, construction and the like. Such planning usually follows the project cycle of identification, preparation, implementation and evaluation. Preparation and implementation are usually clearly differentiated and carried out by different people.

In extension, however, the present is often inadequately known due to lack of intelligence about farmers and their conditions; the future is difficult to specify also, while we do not control events sufficiently to achieve a knowable future. Even if we could remove these bottlenecks, it is still questionable whether blueprint planning could be used in extension. The reason for this is that the target of extension does not respond to laws, it is reactive in that it has its own intentionality. For example, the participation of rural people in programming and implementation has been shown to be a crucial factor for guaranteeing self-sustaining rural development. Participation does not go with blueprint planning.

Process planning is interactive and strategic and based on intelligence; it uses systematic procedures for eliminating unknowns and for identifying activities; it allows for account to be taken of claims from below, participation and information; and it allows organic and incremental development from a small start. Allocation of means is flexible. It is the process that is planned, not its outcomes. Objectives are not formulated in terms of targets but in terms of process functions over time. Process planning seems more suited to extension than blueprint planning.

KNOWLEDGE SYSTEM MANAGEMENT

Extension must be seen as part of a knowledge system in which such functions as knowledge generation, transformation, transfer, utilization and feedback work synergically. In a knowledge system, top-down and bottom-up information link intervening parties and utilizers in a knowledge cycle. As we study them in greater detail, more becomes known about knowledge systems. For example, we now know that successful agricultural knowledge systems are not only characterized by strong intervention power and a good calibration of the research/ practice continuum (good agricultural research, experimental field work, strong linkage mechanisms between research and extension, and effective extension services), but also by strong user control (impact of the utilizer sub-system on research and extension through lobbying, political influence, participation in programming, intelligence gathering and so forth). Intervening parties must be responsive to be effective.

Effective extension seems more and more dependent on effective knowledge system management: monitoring linkages, transformation processes and interfaces between sub-systems, diagnosing bottlenecks in the system, and taking action based upon this knowledge.

Setting extension objectives, whether targeted or flexible, is a function of knowledge system management. Given the perspective of the total knowledge system, it becomes easier to specify the objectives of sub-systems such as extension.

Discussion

The Discussion Groups concerned with the relationship between policy and extension work and the planning of extension activities devoted much attention to clarifying some of the basic principles and concepts involved. These referred particularly to the goals or purposes of extension organizations, the form of the organizations and their management, and to the relationships between these and the effectiveness of extension work.

'Rural extension' was recognized to be a broader concept than 'agricultural extension', although a prime aim for both must be to bring about change. The nature of the change being sought, however, is controversial, opinions differing according to various ideological and disciplinary perspectives. Even so, there was general agreement on several basic considerations. First, there is a direct relationship between the kind of change being proposed and the problems of any particular group of rural people; since all rural people do not have the same problems, the changes to be proposed also have to be different. Secondly, there is a need to distinguish between change which is essentially short-term and linked to technological improvements, and that which is longer-term and linked to social and political trans-formation; it is questionable whether rural extension can legitimately contemplate promoting both forms of change. Thirdly, there are dangers, in terms of possible consequences, of limiting the kinds of change which rural extension can promote too closely to agricultural production. Finally, it is important to recognize different levels of change (e.g., organizational or personal), which require different roles from rural extension workers.

In rural extension, a broad range of activities is involved, and its aims are thus likely to differ from country to country. Moreover, much of this work is being undertaken by a large variety of non-governmental, voluntary organizations. Those whose experience lies entirely with government-operated extension services often find difficulty in recognizing and accepting such organizations. Yet, in many parts of the world they are the only real contact between the majority of the rural people and some form of external support.

For agricultural extension, the main providers are generally formal agencies, usually government extension services. Although the context and content of policy, and thus the emphases in their work, may differ between countries, the prime goal is invariably to assist in realizing increases or improvements in production. Thus, in the main, their work can properly be the provision of relevant knowledge and information in order to stimulate the modernization of agriculture.

The capacity of extension services to be more effective is affected by a number of conditions and constraints. Field extension workers are often conscious that their social status is relatively low, a fact which often adversely affects their performance. However, of greater significance, the effectiveness of extension services is also determined by the manner in which policy is translated into practice. Frequently, the financial resources provided for extension work are inadequate, and the level of activity can become rapidly curtailed if the available government funds are reduced. In addition, many demands are often placed, somewhat arbitrarily, on extension workers to provide various other services to farmers. Although it is recognized that these may be essential for agricultural development, it is arguable whether they should be provided in combination with information; for example, in India, there is a move away from requiring extension workers to adminster subsidies since this is acknowledged as being damaging to the extension work. Further, the effectiveness of extension services is related to the style of management. Constraints are inevitably placed upon this when an extension service is part of a formal bureaucracy, but an autocratic management system, it was agreed, is inimical to effective extension work. Such constraints reflect a lack of knowledge and appreciation among politicians and policy makers of the nature of extension work, of the training requirements of extension personnel at different levels, and of the critical role of leadership in the management of extension services.

One of the main requirements for improving the effectiveness of extension services is coordination, which is primarily a function of management. Attention was drawn to the need to assure good coordination in three particular areas: (a) between extension work and policy, with the need to consider how those involved in extension could influence policy — especially when politicians and policy makers, apart from being relatively uninformed about extension, often are concerned solely with short-term results; (b) between extension and research, without institutionalizing the linkage, (such as occurs, for example, by creating specifically designated liaison officers): in this context, some argued that there is often the need for greater resources to be devoted to research than to extension services; and (c) between extension and the range of services which extension workers may be required to convey to farmers. The last, in particular, bears on the overall role of extension personnel (on which there was no general agreement), especially whether they should be solely concerned with the dissemination of knowledge and information, and, even if this were so, how wide should the range of this educational function be (for example, should it include giving assistance with the establishment of cooperatives).

It was agreed that flexibility in the operational objectives of extension work was preferable to setting rigid targets. Flexibility implies a style of extension management which can respond to the needs of farmers, i.e., be 'demand-led'. To be able to achieve this, the need is for the activities of extension services to be conceived of in terms of a model of a knowledge system. If widely known and understood, this would overcome several of the basic reasons which lead to setting fixed targets. These include: the lack of any practical alternative to target setting; the lack of an appreciation, among planners, of the extension process; the lack of understanding of the functions of knowledge in the system; and the fear that any change from a target-oriented approach, not based on acceptable criteria, would be declared 'irresponsible' by others. However, it was recognized that the acceptance of a knowledge system as a basic model could not, of itself, change other reasons for inflexible target-setting such as, for example: the conventions by which a bureaucratic system functions; the ease of making plans without reference to reality, and the satisfaction of being able to report that targets have been met (even when they have not); and the fear of allowing rural people a voice.

However, despite these limitations, it was thought that basing objectives on a knowledge system model would be useful. It would help in identifying the elements of such a system, in specifying the function of each within the whole, and indicate possible interactions. This would have several desirable consequences. It would enable 'people' (a preferable term to 'planners' since it emphasizes that any member of the system can benefit from a knowledge and understanding of it):

- to compare the real situation with the model and thus identify weaknesses or missing elements;
- to direct information to the most appropriate part or parts of the system;
- to manage their informational needs more effectively, since all participants in the system have to have these needs satisfied if it (the system) is to work efficiently;
- to select communication/information strategies that are congruent with the real knowledge system;
- to identify those parts of the system where information might be distorted or eliminated;
- to understand where they, as individuals, fit in the total knowledge system, and their function within it.

Thus, the model can be used as a political and managerial instrument. Politically, it could be used by members lower down in the system to influence the policies of an organization and the decisions taken by the 'management', while those at the top of the system could use it to influence the policies of national and international agencies. Its main function, however, would be as a managerial tool — to ensure that the knowledge/communication system was complete and active. It was agreed that in such a system a strong feedback loop is essential, and some argued that information pathways from the bottom to the top should be seen by managers as of equal, if not more, importance than those from the top downwards. They argued that only through the rapid tranmission of reliable information upwards could managers be made fully aware of what is happening in the system, on which basis alone could they make appropriate and timely management decisions.

There was general acceptance that the main problem in a knowledge system was information *transformation* (rather than information transfer). This means that the original messages or items of information are altered in their passage through the system, due to two main reasons: (a) the long chains of separate parts, with the possibility of message distortion, amendment, or omission at each link; and (b) information being transmitted that does not fulfil the needs of those at the next level in the system, with the result that they either fail to pass on the information, or, having paid it scant attention, transmit a distorted message. This latter is a serious problem in the transmission of information from policy makers to field staff, but it is even more so in the case of information being transmitted upwards through the system. A full understanding of the system could enable such transformation problems to be overcome.

EXTENSION STRATEGIES AND APPROACHES

8

Strategies in the Transfer of Agricultural Technology, with Reference to Northern Europe

E. DEXTER

ADAS Extension Development Unit, Reading, UK

BASIC ISSUES

Technology Cycle

Figure 1 illustrates elements involved in the transfer of technology.

Technology, defined in this context as 'a way of doing something', can be *generated* by research and by inventive farmers and others. Once a technology has been identified as being promising, it needs *testing* to make sure that it is suitable for the circumstances in which it is likely to be used. If it passes this test it then becomes *proven technology* which can be advocated with confidence through *extension* work.

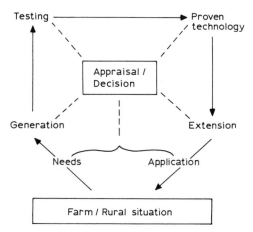

Fig. 1. Elements in the transfer of technology.

The cycle suggests a number of questions. For example, in generating technology, do we make good use of the natural inventiveness inherent in any farming community; who should do the testing — researchers who generate the technology or extension workers who will be advocating its use; and what part should farmers play in the testing of technology and in its subsequent dissemination? But the key elements lie in the processes of appraisal and decision. The key issue, dominating all others, is: who should carry out the appraisals and make the decisions? Other questions which determine strategies are what kinds of agency should be involved in technology transfer; who pays for them; and how can they best achieve the transfer of technology? This paper is concerned with the last issue, specifically with reference to agriculture in some countries of northern Europe.

Extension
A basic function of extension is to assist the transfer of agricultural technology by ensuring that an adequate amount of high quality knowledge about it is present in the farming community. Helping farmers acquire this knowledge involves activity which needs setting in motion. Depending on who it is who takes this initiative, there are three approaches to extension work.

Consultancy
This approach involves providing advice on request. The initiative lies with farmers; they call on extension for a purpose which is determined by them.

Promotional
In this case the initiative lies with the extension agency. Account may be taken of the views of farmers, but decisions on what to do and how to do it are made by the agency.

Participative
This approach is a partnership between farmers and the extension agency and the initiative is shared between them. Together they decide what technologies are important, what information is required and how it should be provided.

These approaches are not mutually exclusive and can be used in a complementary way. To a large extent, the broad strategy of an agricultural extension service is determined by the relative emphasis which is given to them.

Whatever the emphasis, extension work seeks to help farmers make sound decisions on technology by providing information and guidance about it. In so doing extension recognizes: the importance of creating awareness and interest in improved technology, and the need to then help farmers obtain the more detailed guidance they require to assess and apply it successfully; that in any community some farmers are more receptive and responsive than others and adopt technology more quickly; and that people learn from each other — that early adopters can act as foci from which information and technology can spread to other members of the community. Extension activities are concerned with guiding and accelerating these natural processes of adoption and diffusion.

Growing recognition is now being given to the importance of differences in the resources and opportunities available to farmers. Because of these differences, indiscriminate general attempts to accelerate the process of diffusion can have serious weaknesses. Technology which is excellent under some circumstances is inappropriate or irrelevant under others. Therefore, extension work increasingly is being guided by the fact that an agriculture is made up of different sectors with different resources, opportunities and requirements, and that extension (and research) needs to respond to the particular circumstances of each sector.

Extension strategies are based on appraisals of the various agricultural sectors and of the extension resources available to work with them. These appraisals help to answer the questions: what information and guidance do farmers want and need, and how can extension best provide it? The answers determine the nature of the communication linkage between extension agencies and farmers. They also determine the knowledge and skills extension agents need for their work. This, in turn, should guide the information and training they receive, training both in agricultural technology and in its communication.

A wide range of communication methods can be used in the transfer of technology, all of them having their own capabilities and limitations. Used in combination they can be complementary, each one building on information provided by the others. The selection of those to use in any particular situation is based on their ability to give the right kind of information to the right people in the right way at the right time. Communication methods are the tools of the trade of agencies concerned with extension work.

Agencies and Finance

Many kinds of agency are involved in the transfer of agricultural technology. The following are examples:

- Farm input agencies (manufacturers' and merchants' representatives). Their cost is covered by charges made for the goods they supply to farmers.
- State extension services. These are staffed and managed by employees of the government, and are normally financed by state funds obtained from taxes, or by overseas aid. In some cases part of the cost may be met by charging farmers for specific services.
- Local government/community extension agents. These are employed by the community to develop local agriculture or the community as a whole. They may be paid for by local taxes.
- Commodity based agencies. These can take several forms, including farmer-owned and operated cooperatives concerned with production and marketing, which are financed by a levy on sales; quasi governmental marketing boards which can be financed by a levy on sales and charges for services; commodity associations organized by farmers and financed by some combination of membership fees, charges for services and state subsidy.
- Farmer organization based agencies. These are financed by membership subscriptions, charges for services and, in some cases, government subsidy.
- Private consultants. These can be of a business management type or concerned with specific crop or livestock enterprises. They are financed by charging for the services they provide.
- Voluntary agencies financed by aid from charities and other altruistic sources.

Types of Agriculture

Extension deals with widely different kinds of agriculture in which there is an almost infinite number of variations. For the sake of convenience they can be divided into three types:

- Commercial agricultural industries enjoying the benefits of a developed market economy, with its effective marketing channels and effective provision of farming inputs.
- Subsistence agriculture concerned with meeting the basic needs of farming families. The lack of provision for marketing and farm

inputs are two of the ways in which it often differs from commercial agriculture.

— Intermediate agriculture, embracing the many kinds of farming which lie on the continuum between subsistence and commercial agricultures.

COMMERCIAL AGRICULTURE

Although the agriculture of all the countries of northern Europe is commercial, there are striking differences between them in their systems of agricultural extension. The organizations involved have followed different evolutionary pathways and have been shaped by different cultural and historical influences. In order to restrict the scope of this review, examples will be drawn from the countries of Scandinavia and the United Kingdom.

England and Wales
The following are important agencies involved in the transfer of technology in England and Wales:

— ADAS — the Agricultural Development and Advisory Service. This is a state extension service. Its work covers the testing and extension aspects of technology transfer for the whole range of farming systems in the country. [In Northern Ireland there is a separate state extension service, and in Scotland it is university based.]
— Farm input agencies (manufacturers and merchants). These promote the use of their farm inputs and also provide detailed advice on the use of the materials they sell.
— Commodity based organizations set up by statute — for example, the Milk Marketing Board and the British Sugar Corporation. These carry out promotional work and provide advice on matters concerning their particular commodities.
— Private consultants such as crop consultants. In addition to providing advice they arrange the purchase of inputs, often at advantageous prices.

The farm input firms and ADAS are the major agencies involved in technology transfer, most of the day-to-day advice to farmers being provided by the representatives of the manufacturers and merchants.

The strength of ADAS is that it provides an unbiased service based on an excellent foundation of experimental and testing work. Its roles include solving problems other agencies cannot deal with; providing a well-informed and unbiased second opinion on advice provided by other agencies; and providing farmers and all agencies with scientifically sound and up-to-date information on technology. Thus, ADAS exerts a considerable influence on and through other agencies. This magnifies the ADAS input and its contribution to technology transfer. Other important roles are to provide farmers with long-term policy/ management advice on enterprises and systems, and to speed up the rate of adoption of improved technology by those farmers to whom it is relevant.

A strategy used, well suited to speeding up the rate of adoption of technology, involves the combined use of the three extension approaches referred to earlier. The starting point is a small participative group of receptive farmers with a shared interest in a commodity. Working with the extension agent, they adopt improved technology relevant to their farming system. In so doing they may draw directly on the results of experimental and testing work and may be involved in it themselves. This participative work establishes improved technology in the locality. Its wider use can then be stimulated by promotional work which can be guided by the farmer members of the small participative group. This promotional effort stimulates requests for consultancy work to provide specific guidance on the use of the technology on individual farms.

Apart from charges made for certain analytical services ADAS advice has so far been free, but a system of charging more widely for advice is now to be introduced.

Scandinavia
In Finland there is no state run extension service; extension agencies belong to farmers and are governed by them. The following types of organization are involved:

— Agricultural Centres in each of the 15 provinces which carry out general agricultural extension work throughout the country.
— Agricultural Associations, of which there are 14, giving advice on specific branches of farming (for example, poultry and horticulture).
— Cooperatives.
— Local government which, in some cases, employs local extension workers in a socio-economic or community development capacity.

The agricultural centres and associations are financed by a combination of membership subscriptions, charges for services and government subsidy. It may argued that there are too many agencies which could give rise to a duplication of effort. It can also be argued that with this system, central government can readily influence the extension effort through varying its subsidies.

In Denmark also there is no state run extension service, the system being based on extension workers employed by the local farmers' unions, with farmers meeting over 75 per cent of costs. In both Norway and Sweden there is a state-financed extension service, but with all levels being governed by farmers. The one in Sweden has now begun to charge for its services. In Sweden, cooperatives and farmer societies also are important extension agencies. Thus, in Scandinavia, with its wide variations in organizations involved in extension, there is one common theme. They are all governed by farmers and decisions on extension matters are made jointly with their extension workers.

General
There are several important issues which apply generally to the technology transfer systems of Scandinavia and the United Kingdom. First, farmers increasingly are being expected to meet part of the cost of extension work. Secondly, in each country there are groups of specialists whose purpose is to act as a link between local extension agents and the body of knowledge generated by research and testing work. The main role of the specialists is to summarize, interpret and organize the information for use by local extension workers. The specialists also communicate directly with farmers, either on an individual consultancy basis or through extension media. Farmers value the high level of expertise which is available to them in these ways. Finally, recent developments are providing powerful new methods of communication such as video and information technology which will increasingly play an important part in technology transfer. Information technology is based on telecommunication access to a central computer. It is a two-way system of communication with the ability to provide immediate access to the latest and best information, and to use that information to recommend optimum decisions for particular farms and fields. Equipped with a terminal, any extension worker, farmer or community will have direct and immediate access to the best knowledge and guidance in the land. The computer will then take the place of the oracle.

9

Strategies in the Transfer of Technology: The IBFEP Approach

S. P. DHUA

Hindustan Insecticides Ltd, New Delhi, India

BACKGROUND

In recent years, India has made remarkable progress in the field of agriculture. The first few years of the post-independence era saw a near stagnant agriculture. The strategies taken in the Second Five Year Plan started yielding good results and the country witnessed a steep rise in agricultural production. In the year 1983–84, the foodgrain production touched a new high with a production figure of 151·6 million tonnes. By the turn of the century, the country should, and is expected to, produce around 220 million tonnes of foodgrains.

The key factors contributing to the success of the new agricultural strategy in the mid-sixties were the development of new high-yielding varieties of crops and the transfer of an improved package of cultivation technology to the farmers. The decade of the seventies saw the implementation of crop-specific programmes and special programmes for small and marginal farmers (e.g., the Small Farmers Development Agency and Marginal Farmers Development Agency). In the second half of the seventies, a massive extension programme was launched in different States of India under the financing and guidance of the World Bank. This was the Training and Visit (T & V) programme, which was expected to accelerate agricultural productivity and production. With the multiplier effect that it was to generate, one would expect a high rate of growth of agricultural production.

While this was the overall scene, the states of Eastern India faced peculiar problems. Overall, these states still have low productivity, a low and unbalanced usage of fertilizers, the majority of farmers are small and marginal, and there is a predominance of poor peasantry.

THE INDO-GERMAN FERTILIZER EDUCATION PROJECT

The scene needed to be changed radically. The Hindustan Fertilizer Corporation (HFC) started by having extensive fertilizer education programmes spread over three or four Community Development Blocks for each extension worker. By the early seventies, HFC realized that an intensive programme, dovetailed with an extensive programme, could achieve better results. With that aim in view, HFC launched the Indo-German Fertilizer Education Project (IGFEP) financed by the Government of the Federal Republic of Germany in the State of West Bengal. The central theme of the Project was to move sound agricultural practices to the farmers. An agricultural graduate was posted and stayed in a key village where he was to undertake crop improvement programmes for the farmers of the village. In the surrounding villages, extensive promotional campaigns were undertaken. The favourable impact of the Project, as evaluated by the National Council of Applied Economic Research, New Delhi (NCAER) in 1976, was evident in the movement towards greater cropping intensity; the increased area being fertilized; extensive use of high-yielding varieties of improved seeds; an increased level of fertilizer consumption; better nutrient balance; and improved yields.

In order to measure the welfare or distributional impact of the Project on the population covered, a second study was entrusted to NCAER in 1980 with the objective of measuring the economic impact of the Project on small and marginal farmers and landless workers in West Bengal. This study showed that considerable benefits were being drawn by the small and marginal farmers from the Project.

THE NEW APPROACH

The experience gained in implementing the IGFEP showed that besides the knowledge input, the whole series in a service delivery system connected with the supply of requisite inputs to farmers in a sequential order, easily, adequately and in a timely manner, markedly influences the success of crop production as well as productivity. Unless the knowledge is harmonized with the requirements for input arrangements and associated farm services, efforts in teaching and motivating farmers on improved technology become futile. Any shortcoming in the intermediate stages of the input delivery system causes aberrations in the food production chain.

The experience of the T & V system with its major thrust on the knowledge input but not backed by an input delivery system was also not encouraging. Esman and Uphoff (1984) have given as their opinion that 'scepticism about inefficiency of the system has mounted, however, precisely because no group basis for communication and innovation was established, and the contact farmer's linkage to other farmers has been weak'.

The intense desire to help farmers with the knowledge input as the first step in transferring technology to the market place, and then to arrange all the required inputs of the farmers in a systematic manner, led to the implementation of the Indo-British Fertilizer Education Project (IBFEP), financed by the United Kingdom Government in the six States of Eastern India, viz. Assam, Bihar, Madhya Pradesh, Orissa, Uttar Pradesh and West Bengal. The Project was not conceived as a mere extension tool, but as a philosophy. In terms of theory, the Project was fairly close to the T & V system in the form of the knowledge input but differed significantly in its thrust on providing all the inputs to the farmers. The input requirements of the farmers were to be closely monitored, and assistance given to the farmers to obtain the supplies in time. Like the T & V system, a very high priority was given to the training of all the field staff on the latest technology available. Agronomists, soil scientists and plant protection staff were given training to keep them abreast with the latest research work in related fields. The Project aimed not only at working with the farmers at the market place but also to bringing research right to their doorstep.

THE INDO-BRITISH FERTILIZER EDUCATION PROJECT

The objective of the present paper is to share our experience of implementing the Indo-British Fertilizer Education Project. This dynamic model, which the IBFEP provides, the author hopes will be tried and replicated in different countries with similar problems so that the beneficial impact can be used to increase agricultural production.

The Project
In view of the slow process of agricultural development in the Eastern states, the Hindustan Fertilizer Corporation Ltd has been executing the IBFEP in 25 selected districts of Assam, Bihar, Orissa, Madhya Pradesh, Uttar Pradesh and West Bengal since the 1981–82 *rabi* season. The Project aims at increased food production in the six states by

educating farmers and demonstrating in their fields the profitability of scientific fertilizer usage and adopting an improved package of practices, and also concurrently ensuring the timely availability of inputs. The Project is financed by the United Kingdom Government for a period of five years under a bilateral agreement between the Government of India and that of the United Kingdom. Special emphasis has been attached under the Project programme to benefit the majority of small and marginal farmers.

With a view to exploiting the production potential, districts with a fairly good irrigation potential have been selected by the Government of India in consultation with the concerned State Governments. Nevertheless, some backward districts warranting immediate attention have also been included under the Project programme. In each of the selected districts, four Community Development Blocks, and within each block, two clusters of villages, each cluster consisting of ten villages, have been taken for the fertilizer education programme. The Project thus covers 2000 villages spreading over 100 blocks in 25 districts. A cluster of 10 villages constitutes the operational unit for one experienced agricultural graduate, designated as 'Cluster Agronomist', who is resident in one of the villages of the cluster. The Cluster Agronomist is the key figure of the Project and he is responsible for undertaking a large-scale demonstration spread over two villages for one year, supplemented with other extension aids, and more importantly, ensuring that the farmers receive the required inputs. In addition to this grassroots level staff, an Agronomist posted at block level, besides guiding the implementation programme of the Cluster Agronomists, also undertakes the follow-up activities in post-demonstration villages when the demonstration site is moved to other villages. The activities of eight clusters of villages in a district are supervised by a District Agronomist, posted at District Headquarters, and the total activities of the State are supervised and monitored by the Regional Project Leader, posted at State Headquarters.

For proper planning and implementation of the IBFEP's action plan, the Project coordinates very closely with the State Government Department of Agriculture through duly constituted State, District and Village Level Coordination Committees under the Chairmanship of Agricultural Production Commissioner, District Magistrate and Block Development Officer, respectively. At the apex level, the Project is administered through an All India Level Committee with the representatives of the concerned Ministries of the Government of India and the Agricultural Production Commissioners of the concerned States. In order to obtain a comprehensive idea on the impact of the Project, the National Council of Applied Economics Research along with Projects and Development India Ltd have been entrusted to undertake a periodic assessment and to suggest remedial measures from time to time.

Action Programme

In order to achieve the prime objective of the Project in educating farmers on scientific fertilizer usage along with an improved package of practices, and simultaneously ensuring the supply of requisite inputs for an effective translation of knowledge input, the following action programmes are being implemented:

(1) Establishment of a large-scale block demonstration of about 60 ha area (in two villages) for one year.

(2) Mobilization of organizational resources in ensuring the arrangement of production inputs and associated services to the farmers in an integrated manner.

(3) Arrangement of free soil testing and a fertilizer recommendation service.

(4) Supply of fertilizers and plant protection chemicals at subsidized rates to encourage the participation of small and marginal farmers and also reduce the risks.

(5) Reinforcement of demonstrations with adequate supplementary extension activities (such as group meetings, farmers' training, field days, crop seminars and workshops, farmers' tours, distribution of technical literature, etc.)

(6) Augmentation of the infrastructural base for inputs supply and integrated agricultural development (such as additional irrigation facilities, seed multiplication and exchange, land development, bio-gas plant, etc.)

The main extension thrust of the Project for the transfer of technology to farmers' fields has been the implementation of large-scale demonstrations covering a large group of farmers. In order to encourage the participation of small and marginal farmers in the programme, a 50 per cent subsidy on fertilizers and a 30 per cent subsidy on plant protection chemicals are allowed for one year with the expectation that the additional income generated from higher yields in demonstration would be recycled in furthering agricultural development. In order to render more benefit to small and marginal farmers through the Project, the fertilizer subsidy is restricted to 1 ha area only for medium and big farmers, but not for small and marginal farmer categories. In order to ensure a better economy in production, farmers are guided to apply fertilizer in conjunction with organic manures based on soil analysis of individual plots. Based on the soil test value, the Agronomist issues a delivery order in the name of the participating farmer to procure fertilizer at half price from the nearest retailers. The soil testing service of the Project is made available to the farmers through static and mobile soil testing laboratories established at district level. After the selection of demonstration areas, the listing of participating farmers, the issue of fertilizer and the mobilization of inputs (viz. seeds, credit, plant protection chemicals, etc.), Cluster Agronomists make an all-out effort in order to assure the success of each block demonstration.

The Impact of the Project

As a result of the IBFEP activities, fertilizer consumption in the Project areas has increased by about 45–60 per cent, besides significant improvement in the balanced use of plant nutrients based on soil testing. The area under the high-yielding varieties of crops has also risen by about 50–60 per cent. The yields of crops have shown significant increases in the Demonstration Blocks; 61 per cent in the case of paddy, 80 per cent in the case of wheat, 40 per cent in the case of maize, 88 per cent in the case of *jowar*, and 75 per cent in the case of mustard. Consequently, the average net profit of the farmers participating in the programme has increased by about Rs. 1200–1500 per hectare.

This impact is of significance since 70 per cent of the participating farmers were small and marginal, and the main focus of the Project was on this section of the farming community.

PROPOSED FUTURE STRATEGY

Based on the IBFEP experience, the following strategy is suggested for undertaking a worthwhile extension education programme in developing countries:

(i) An extension programme at the grassroots level must be adequately supported with a sound input delivery system.

(ii) The placement of a trained person, preferably an agricultural graduate, at the village level for effectively administering the supply of agricultural inputs along with technology transfer at a significantly high level, primarily amongst the small and marginal farmers, is indispensable.

(iii) A Block Demonstration approach ensures the involvement of the majority of small and marginal farmers and creates a visual impact which is vital for achieving a 'ripple effect' in the adjoining areas.

(iv) For any fertilizer programme, an Integrated Plant Nutrition System should be followed based on soil testing rather than specific crop-related fertilizer recommendations.

(v) To guard against the uncertainties of production factors, the introduction of a Crop Insurance Policy is essential where large numbers of small and marginal farmers are to be covered.

(vi) Adequate credit support for helping the small and marginal farmers to buy inputs is indispensable.

(vii) The development of a suitable dry land farming strategy for relevant areas is essential.

10

Strategies in the Transfer of Technology

PETER W. C. HOARE

Highland Agricultural and Social Development Project, Chiangmai, Thailand

INTRODUCTION

The transfer of technology in agricultural development has generally been directed towards increased physical production from farms, and reducing costs to farmers (Gittinger, 1982, page 56). During the last decade there have been policy commitments by international and bilateral development agencies to the development of the rural poor (World Bank, 1975, page 4; AOAP, 1984, page 59). The origin of this concern for the rural poor stemmed in part from the realization that the traditional transfer to technology processes were mainly benefitting those farmers with superior resources, and widening the 'equity gap' (Crouch and Chamala, 1981*b*, page 233).

This paper will start with the traditional transfer of technology (TOT) model, and then examine the emerging strategies for both technology generation designed to benefit the rural poor and poverty oriented extension. 'Spontaneous' adoption of new technology by both large-scale and small-scale farmers may take place without government intervention, when agri-support services are well developed and favourable price relations for agricultural commodities exist. One such example is the increase in Thai exports of tapioca products from 1·9 million tons in 1978 to 6 million tons in 1982.

Strategies for generation of new agricultural technology are changing in response to the non-adoption of new technology by farmers with limited resources in developing countries. Chambers and Ghildyal have recently summarized the dominant models in the post-war years, and proposed a new model to focus on the special needs of 'resource

137

poor farmers', i.e. those 'whose resources of land, water, labour and capital do not currently permit a decent and secure livelihood' (Chambers and Ghildyal, 1985, page 3) (see Table 1).

TABLE 1
Non-adoption of Technology by Small-scale Farmers: Changes in Explanation and Prescription

Model	Period when dominant	Explanation of non-adoption of technology	Prescription for adoption of technology
TOT	1950s 1960s	Ignorance of farmers	Agricultural extension to transfer the technology
TOT (modified)	1970s 1980s	Farm-level constraints	Ease constraints to enable farmers to adopt the technology
FFL	Latter 1980s for RPFs?	The technology does not fit RPF conditions	FFL to generate technology which does fit RPF conditions

TOT = Transfer of technology; FFL = Farmer-first-and-last;
RPF = Resource poor farmers.
(Source: Chambers and Ghildyal, 1985, page 22).
Notes
(1) TOT is the 'traditional' model based on the development of new agricultural technologies on research stations, followed by the transfer of this technology to farmers.

(2) TOT (modified) is the model developed by research scientists at IRRI, CIMMYT, ICRAF, CIP, Sondeo and other locations in response to the failure of farmers to adopt new technology developed on research stations. These strategies of technology development involve the use of multidisciplinary teams of scientists to define the resource constraints under which farmers operate, and to conduct on-farm experiments on farmers' fields to develop new technology.

(3) FFL is a model proposed by Chambers and Ghildyal which entails a fundamental reversal of learning and research strategy: 'Research problems and priorities are identified by the needs and opportunities of the farm families rather than by the professional preferences of the scientist' (Chambers and Ghildyal, 1985, page 13).

THE TRANSFER OF TECHNOLOGY (TOT) MODEL, AND AGRI-SUPPORT SERVICES

The TOT model is described in some detail to highlight the role of agri-support services in the transfer of technology. Mosher (1978) illustrated this concept of technology transfer based on what he called the 'achievement distribution' of farmers. If the assumption were made that farmers in a particular locality had similar resources of land, labour and capital, then some farmers would achieve higher production than others, and the actual production achieved by the best farmers never equalled the production that was possible from the application of the best farm management skills for that locality (Mosher, 1978). When the farmers were ranked in terms of value added per unit area on their farms (i.e. the combined value of all the products produced minus the combined value of all the resources used in producing these products), then the distribution pattern shown in Fig. 1 was obtained.

The comparison of the technical and economic ceilings in Fig. 1a is based on the physical production per unit area (Scale B). The technical

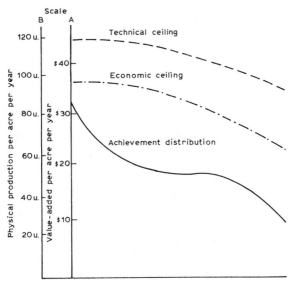

Fig. 1a. Relationship of technical ceiling to economic ceiling and achievement distribution. (Source: Mosher, 1978, page 73.)

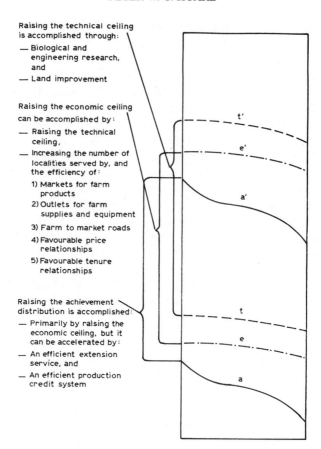

Fig. 1b. Main means available for raising the technical and economic ceilings and the achievement distribution. (Source: Mosher, 1978, page 68.)

ceiling, which represents what can be accomplished through biological and engineering research and land improvement, is always above the economic ceiling. On the other hand, the comparison between the farmers achievement distribution and the economic ceiling is in terms of value added per unit area per year (Scale A). The raising of the economic ceiling concerns changes to the agri-support services. These services are (1) markets for farm products, (2) outlets for farm supplies and equipment, (3) farm to market roads, (4) favourable price relationships, and (5) favourable land tenure relationships. The raising of the achievement distribution of farmers is dependent mainly on raising the economic ceiling, and this process can be accelerated through an efficient extension service, and an efficient production credit system. The two basic strategies for raising farm productivity in this model for technology transfer are first, to raise the technical and economic ceilings, and/or, secondly, to raise the achievement levels of the farmers at the low end of the distribution curve to approach the high achievers. However, this later strategy has limitations. Where all households start with a similar land resource, within one generation about one third of the members of a rural community will be at the low end of the 'achievement distribution' due to factors such as poor health and various social reasons.

The Indonesian paddy rice production intensification scheme provides a successful example of the application of this model of technology transfer in a developing country. The agricultural development strategy used in turning Indonesia from the world's biggest rice importer into a rice exporter included: (1) the development of high yielding rice varieties; (2) improved irrigation infrastructure; (3) increasing farmer participation in small-scale irrigation development; (4) a national policy to encourage fertilizer application by subsidizing more than half the cost; and (5) an effective extension service and agricultural credit supply to help raise farmer achievement based on growth, equity, and stability (Sanusi, 1983).

However, reviews of the experience of the past three decades in rural development projects in developing countries have shown that few if any of the projects have accomplished the goals set forth in the project design phase (Evenson, 1984). Those who have benefited least from intervention are the resource poor families, who in 1975 numbered over 500 million.* One reason for the non-adoption of new technology by resource poor farmers is that many view profitability in terms of return per labour day worked, whereas the technology developed by research scientists is usually based on increasing yield per unit area (van de Laar, 1982, page 148; Hoare, 1984a, page 194). Research scientists at international research institutes are now giving more attention to technology development within farmers' resource constraints.

---

*This figure was based on the World Bank's estimate (World Bank, 1975, page 4) that 85 per cent of the then 750 million rural poor in the developing world were considered to be living in absolute poverty, with a per capita annual income of less than US $50.

THE MODIFIED TRANSFER OF TECHNOLOGY MODEL*

The research scientists of the International Rice Research Institute (IRRI) noted that when research workers were obtaining irrigated rice yields of 6 to 8 tonnes per hectare, competent farmers were obtaining only 3 to 4 tonnes per hectare, and many were 'lucky to get 2 t/ha' (De Datta *et al.*, 1978, page 3). These observations led them to examine the resource constraints under which small-scale farmers operate their farm enterprises. In the model which they prepared, in order to explain the difference between the experimental station and the farmer's field, there were two 'yield gaps'. The first and smallest 'yield gap' was the environmental difference between the experimental station and the farmer's field (non-transferable) (Fig. 2). The second and larger 'yield gap' was the difference between the potential and actual yield obtained

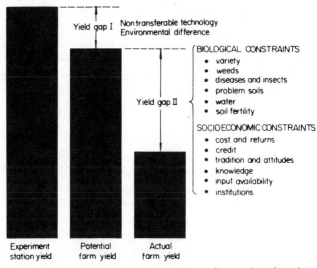

Fig. 2. The concept of yield gaps between an experimental station rice yield, the potential farm yield, and the actual farm yield. (Source: De Datta *et al.*, 1978, page 4.) (Reproduced by permission of the International Rice Research Institute.)

*Other prototypes of the transfer of technology model (TOT) have been developed. These include: (1) the CIMMYT approach; (2) the Sondeo approach developed by Hildebrand (1981) in Guatemala; (3) ICRAF's diagnosis and design; and (4) CIP's farmer-back-to-farmer research (Rhoades and Booth, 1982).

on farmers' fields. This was due to the farmers using inputs or practices which result in lower yields than are possible. They found that the constraints to maximum production per unit area were biological and socio-economic. This led to a modification of the TOT model with on-farm research taking into account the farmers' resource constraints and recognizing the interrelationships and trade-offs between different farm enterprises.

The common activities in the Farming Systems Research and Development (FSR & D) approaches of IRRI, CIMMYT, ICRAF, CIP and Sondeo are: (1) target and research area selection; (2) problem identification; (3) planning of on-farm research; (4) on-farm research and analysis; and (5) extension of results (Shaner *et al.*, 1982). The starting point is mostly a rapid rural appraisal by multidisciplinary teams of scientists in order to understand the farmers' present utilization of land, family labour, capital, farm management skills; and the small farmers' needs and priorities (Collinson, 1981).

There are problems of interdisciplinary cooperation encountered with teams of research scientists working with on-farm FSR & D. Raintree, working on agroforestry systems at ICRAF, described some of the institutional constraints and communication problems in the transition from a 'mere multidisciplinary to a genuine interdisciplinary approach'. These were largely due to the differences in the interests and perspectives of research scientists (Raintree, 1984). There are also different perspectives between the research scientists and the farmers to consider. Chambers and Ghildyal (1985) have remarked that it is difficult for a social scientist to learn to think like an agronomist, and even more so for either of them to learn to think like a small farmer. These authors also considered that during the design phase of the on-farm research, the priorities of the scientist would take precedence over those of the farm families.

THE FARMER-FIRST-AND-LAST MODEL

The justifications for explicit reference to equity in agricultural technology in the Farmer-First-and-Last (FFL) model described in Table 1 are: (1) social justice, as three-quarters of the farms in India are less than 2 hectares (over 60 million farm families in 1984); (2) the potential for increased agricultural production is great, as 75 per cent of the area in sub-Saharan Africa is cropped by resource poor farmers;

and (3) employment opportunities would increase both on-farm and off-farm. Chambers and Ghildyal envisage the technology generation as a synthesis of the indigenous farmer technology and the technical knowledge developed by research scientists. The institutional arrangements to develop this concept, however, have yet to be worked out.

EXTENSION STRATEGIES FOR THE TRANSFER OF TECHNOLOGY

I will confine the discussion to face-to-face extension contact, which Cernea *et al.* (1983, page 144) consider particularly important in subsistence agriculture, where there are large numbers of illiterate farmers.

Extension Strategies Based on the TOT Model

The World Bank has provided loans for over 40 developing countries in Asia, Africa, Europe, and Central and South America to adopt the Training and Visit extension management system based on the TOT model (Benor *et al.*, 1984, Preface). In this system, the Village Level Extension workers (VEWs) who serve about 1000 farm families are trained fortnightly in the new technology developed by research scientists based on the TOT model. Contact farmers are selected who are expected to apply the new technology in their own fields, and transfer the knowledge to other farmers. The contact farmers would be expected to be those with above average resources, in order to have the maximum impact on increasing agricultural production (Röling, 1982, page 88).

The concensus of the participants at the First Asian Workshop on the T & V System in Chiang Mai in 1981 was that this strategy of technology transfer had benefited fewer farmers in upland rainfed areas. One reason for this is the greater complexity of new technology packages for upland rainfed areas in the wet tropics, for in such areas there is the need to address soil conservation problems, e.g., the causes and extent of soil erosion, cropping systems which will help control soil degradation, and the amelioration of land already degraded (Anecksamphat and Buddee, 1984).

Cernea *et al.* (1983, page 147) advocate a strategy of providing better training for VEWs in farm management, and the sociology of rural communities in order to improve their efficiency in technology transfer. A similar point has been made by Lamrock (1979, page 5) who, after twenty years field experience in Papua New Guinea, considered that VEWs should be trained in agricultural technology, production economics, the behavioural sciences, and the processes of adult education.

Extension Strategies Oriented Towards the Alleviation of Poverty

The FAO Small Farmers' Development Project (SFDP), begun in the early 1970s, is aimed at the resource poor, tenant farmers, the landless, and fishermen in the Asian Region.

In December, 1982, there were 3272 small farmer groups in Nepal, Indonesia, Bangladesh and the Philippines (FAO, 1983). The strategies to benefit the rural poor include: (1) the generation of a receiving/utilizing system of multifunctional groups below the level of existing rural institutions in order to obtain a share of the government inputs for poorer families for development; (2) the formation of small, homogeneous trust groups which have common income generating activity and are capable of using joint liability credit; and (3) group organizers, who are initially required to catalyze group formation.

The technology needs identified by the small farmer groups include improved ways of grain storage, and grain milling, and cooking with limited fuel (FAO, 1983). The development of extension–research linkages for the development of new technology is not central to this FAO/SFDP strategy. However, Garforth (1982b, page 62) considered that poverty-oriented extension work needs to incorporate the following three elements if it is to be productive: (1) meaningful two-way communication between research and extension staff, and extension staff and the rural poor; (2) a willingness to work with the rural poor to clarify problems and identify solutions; and (3) a flexibility in programme objectives and components.

Farmer-Centred Extension Strategies

Farmer-centred extension techniques (FACE) have evolved during more than two decades of Australian experience in rainfed agricultural development in the Asian–Pacific Region (Lamrock, 1966; ADAB, 1981; Tully, 1981; Hoare et al., 1982; Hoare, 1984b). These techniques include the three elements identified by Garforth, and the strategy is now being applied in the implementation of the International Development Association/World Bank funded Highland Agricultural and Social Development Project in a range of irrigated and rainfed environments in North Thailand.

The steps in the development of technical packages, and the extension process are: (1) the problem census meeting to obtain farmer collaboration in the identification and ranking of the main agricultural problems of the community; (2) problem solving meetings with rural communities to explore possible solutions to their agricultural problems; (3) modifications to the macro plan, where necessary, based on the problem census and problem solving meetings; (4) an annual budget planning workshop where agricultural development plans prepared by farmers and VEWs, which are in line with project policy, are allocated budget in the forthcoming financial year; (5) on-farm demonstrations with a set of improved cultural practices which combine indigenous technical knowledge (e.g., farmers' own rice varieties) with an improved package of cultural practices; (6) subsidized inputs and revolving funds to enable resource poor farmers to apply new technical packages at minimum risk; and (7) a sustained training programme for VEWs in the new technology, and the monitoring of VEW to farmer technology transfer.

Farmer adoption levels of over 20 per cent in two years have been achieved for paddy yield increase in part of the project area (Hoare and Wells, 1984). In some small highland villages there has been group adoption by all farmers in the community, within one year, of new technology for tea leaf production (Hoare, 1984b).

SOME KEY ISSUES

(1) Are the expectations for developing appropriate technologies for resource poor farmers attainable? The biological and socio-economic constraints to crop production in upland rainfed areas are greater than in irrigated fields. The technology is also more location specific, and credit is less available.

(2) National extension services usually focus their efforts on the farmers with above average resources in order to achieve their objectives of increasing national food supply and agricultural commodities for export. Are these objectives compatible with raising incomes of resource poor farmers ? Is a separate cadre of rural development workers needed to concentrate solely on the resource poor farmers?

(3) There seems to be a concensus in the literature that the skills of the VEWs and other rural development workers need to be improved.

The new skills needed are in farm development planning and group processes in order to facilitate the transfer of technology. This will require an increased investment in the training of field staff.

ACKNOWLEDGEMENTS

The author acknowledges the opportunity to participate in extension research, education, and training work in North Thailand with staff of the Faculty of Agriculture, Chiang Mai University and the Public Welfare Department of the Kingdom of Thailand; provided through the Australian Development Assistance Bureau with the Thai Australian Highland Agricultural Project, and the Highland Agricultural and Social Development Project. The views presented in this paper are those of the author and not necessarily those of the organizations mentioned.

I also acknowledge the assistance of my colleague Mr Peter Jones in the implementation of the extension training programme of the Highland Agricultural and Social Development Project, a sub-project of the Northern Agricultural Development Project funded through an IDA/World Bank loan in Northern Thailand. The Australian Development Assistance Bureau is providing technical assistance to the Public Welfare Department, Hilltribe Development Division, to assist with the implementation of the project. The Australian Agricultural Consulting and Management Company is managing the Australian advisory team.

11

Planning Extension Programmes Appropriate to Rural Women

JOAN ALLEN PETERS

Department of Home Economics, Bath College of Higher Education, Bath, UK

When you think of extension programmes for women, what comes to mind? A group of smiling women gathered around an extension worker demonstrating how to make a nourishing meal? A young village woman bending over a bit of weaving or basketry for the craft trade? A group of small girls learning to sew? All who have worked with women in extension will have at least one favourite image that instantly springs from memory when asked that question. But as we take a more searching look back, at the end of the women's decade, can we smile with satisfaction at the progress made towards improving the lot of rural women?

In spite of occasional evidence of some success, the plight of most rural women is improved little over what it was ten years ago. Why have so many well-intentioned rural development programmes, including extension efforts, made so little lasting impact on the women they were supposed to benefit?

One of the main reasons for failure, or at best limited success, is that women and their real needs have seldom been actively taken into account when projects are planned. And the reason for this neglect is frequently because women and their work were not included in the data used for project planning.

Do women in the Third World's rural areas work? It is easy for most people to say 'Yes', they do collect firewood and fetch water, often walking many miles to do so. And 'Yes', they also care for children and spend long hours washing clothes and cleaning the home and its immediate surroundings. We may even remember that they tend and water the family's chickens and other animals; and process foodstuffs

for market, and carry the produce to barter or sell in the market place. It
is easier to forget, however, that they also are involved in all aspects of
agriculture. They plant, cultivate, weed, harvest and process the chief
food crops — plantain, rice, maize, cassava, tubers and vegetables. In
Africa, 60–80 per cent of the agricultural labour force is female, as is 40
per cent in Asia and the West Indies. Yet, in spite of these efforts, women
are often not considered as contributors to the family's income, but
rather as merely helpers to their men, or as simply 'doing women's
work', their expected role since time began.

Since they neither own nor control the use of the land they work,
women's efforts are usually economically invisible, not only to their
husbands and families, but to many development agencies. When
projects are planned, women and their work are not included in the
statistics used for planning. The GNP technique, which measures
national output and productivity, omits those informal activities in
which women are particularly involved, such as subsistence food
production, child care and breast-feeding, because these activities have
no monetary value.

Because rural women do not show up in the planning statistics, rural
development and agricultural extension projects are often directly
aimed towards men. In fact, they are usually planned, carried out and
evaluated by men. The introduction of new agricultural practices,
technical assistance, appropriate technology, industrialization and
income-generating projects such as cash cropping, have all suffered in
the past from being male dominated. Consequently, these activities
have sometimes damaged the well-being of women. Instead of closing
the gap between rich and poor, these patriarchal development schemes
widen the gap between women and men still further.

In addition to the kinds of problems that may result from male-
dominated development systems, women also have those particular
problems which evolve through the perennial stress which the unique
roles of childbearing and being primary caretakers of children ascribe
to them biologically and socially. Childbirth is still a major cause of
death in many developing countries, especially among the poorest, most
undernourished sector of the population. Yet in many countries
women's freedom to obtain and use birth control is restricted. As a
result, in many Third World countries, they can still expect an average of
eight pregnancies, resulting in anaemia and nutritional deprivation. Yet
women's work loads, in the home and in the fields, are not lessened
appreciably with the increase in family size.

Thus, sexual discrimination, manifested by unequal access to ownership of land, insensitive development projects and traditional patterns of domestic and social life, continues to be a major factor that denies women proper recognition and access to vital resources, including rural extension programmes. What can we then suggest, from our experience of the past two decades, as guidelines to follow when planning extension programmes for women, in order to ensure their usefulness and relevance to those they are meant to serve?

(1) *The planning of extension programmes must be based on accurate and appropriate data which reflect the true status and needs of women.* The United Nations Development Programme made this recommendation in 1980 when it suggested that systematic studies of women's work should be undertaken not only as part of the technical cooperation projects already assisted by the International Labour Organization and the United Nations, but also through the inter-regional and national household survey capability programme being sponsored by the World Bank, the UNDP, and the UN. At country level, household surveys on a wide range of topics pertaining to the socio-economic conditions of women, including employment, were seen as essential to form a more accurate data base for planning.

The amount of available data on women has increased considerably over the past decade. The State of the World's Women Report (1985) reveals much valuable information for planning, but concludes that national surveys invariably under-estimate women's work, particularly in agriculture. And it is still true that women the world over are making a vast, and generally unacknowledged, contribution to the welfare and wealth of their communities in unpaid household work and in small-scale business and market activities.

It is also true that the data which have been collected are often not made widely available for planning. Relevant case studies and surveys must be disseminated in a form which can be readily understood and used by programme planners. Improvements in communications technology sometimes seem peculiarly ineffective in getting the facts into the field where they are most needed. Planning and implementing groups may also need help to use information in the best way. Women should be encouraged and trained to collect, analyse and use data in their own programme development.

(2) *Women's development must be viewed as part of the development of the whole community, and must be provided with adequate resources at all levels.* The general tendency in most developing countries has been to separate

development into narrow compartments, sometimes entrusted to different administrative agencies or ministries. Women have been included primarily in programmes for health, education and family welfare, but often forgotten in the designing of economic development programmes and institutional infrastructures. Schemes for the location of large projects such as irrigation; income generation programmes; creation of large multipurpose corporations; or improvements to transport or communications systems, all may affect the lives and well-being of women. Such development may increase or even introduce inequalities into the society. When it does, women (at least the poorest ones) become doubly underprivileged, first as members of the poor, and then as women.

A careful examination of women's role at early stages of development schemes, and the creation of built-in systems to ensure involvement of women in all development efforts, can help to avoid the negative effects of development on women. The participation of women, particularly of the actual target groups, is needed (a) to change or reorient thinking at policy-making levels; (b) to move from generalizations to specific measures; and (c) to increase women's self-reliance and determination to play a more active role in the development process.

Resources, both human and material, must be allocated to women's programming. This will happen only when policies which stress the importance of women's programmes are adopted by governments and granted not only lip-service but adequate financing and support.

(3) *Women require training and information to prepare them for their involvement in development.* In some countries it is now understood that development, including women's development, involves the empowerment of those who have been generally deprived of the opportunities to learn decision-making skills and to participate in the making of decisions. Thus, the training needs of women are seen to be more than those in the areas of health and cottage crafts. Training will have to concentrate on the skills needed for motivation, empowerment, leadership and participation. Training programmes have to be designed to facilitate women to experience the concepts of communication and participation, and to become aware of, and comfortable with, the power that should be theirs. Such training also requires skilled 'trainers' who are involved in and committed to the style of participatory learning which permits 'trainees' to influence the training programme and design priorities.

Information too is often not designed for, or aimed at, women. The assumption is that men are the decision makers and income earners in a family, and that any information communicated to them will somehow get through to their wives. One study in an area of Kenya revealed that male farm managers were fourteen times more likely to have been contacted by extension workers than women. Sometimes women are not even aware of the existence of specific services or programmes designed for them.

Thus, women have different and special kinds of information needs, and these will be different at various stages of the life cycle. Because their literacy levels are usually lower than those of men, they will need information in forms that they can understand. Interpersonal and group communications often play a larger role in facilitating the flow of information to, from and among rural women than do the information sources and channels such as written materials, product labelling and mass media. It is vital to assess the patterns of interaction in a community, the attitudes of communities to new ideas and practices, and the local cultural, economic and political conditions. A combination of appropriate methods of mass media, interpersonal and group communications can successfully provide the kind of information needed at the different stages in the educational process, from awareness of the idea or practice through acceptance to adoption.

(4) *Women's programmes must recognize the imbalance which exists in the work loads and working conditions of men and women.* The commitment of governments to improving the position of women has led to world-wide improvements in the education and health care facilities. But the burden of domestic work for most women, often combined with the need to work outside the home, has negated some of the positive effects of these improvements. Women still have less time for study, for taking part in political and social activity, and for making use of available resources and facilities for health and development. Some of the facts contributed by The State of the World's Women (1985) emphasize the inequities which exist between the sexes:

— A housewife will spend on average 56 per cent of her time in unpaid work; a man with a job, 11 per cent. But a woman with a job still spends on average 31 per cent of her time in unpaid domestic work.
— Women in Africa do up to three-quarters of all the agricultural

work in addition to their domestic responsibilities. Women farmers grow half the world's food, but local studies show that national surveys invariably underestimate women's agricultural work (e.g., in Egypt, national figures indicate that women contribute 3·6 per cent of the agricultural work but local figures show the proportion to be closer to 35–50 per cent).
— In the last decade 100 million more women have gone into wage employment, but their proportion of the total work force has risen only about 4 per cent since 1950, and not at all over the Women's Decade.

Many programmes designed for women neglect the already heavy burden of work they carry. For example, income generation or self-help schemes for women are impractical or even harmful if they neglect to provide labour-saving practices or technologies so that women will not merely be adding more hours to their working day, or be forced to neglect vital child-caring and family feeding responsibilities. Programmes and projects which address the problems of providing safe, adequate and accessible water supplies for the household, which reduce the labour of food production and preparation through culturally appropriate technologies, and which look at ways to conserve fuel through exploration and application of alternative energy sources, must be a top priority.

(5) *Women's programmes should provide income to women, and the power which accompanies it.* Activities rewarded financially assume importance in the eyes of the community as well as of the individual. Increasing efforts must be made to place an economic and monetary value on those things which women do as a regular and often unseen part of home and family life. In some cultures men are beginning to recognize what a woman is worth when, for reasons of death, divorce or illness, they have to hire a worker to carry out household activities formerly performed by the partner. Is it not desirable, or at least possible, to consider providing a regular salary or allowance to women for work in the home? Payment may restore to the woman's work the importance and dignity it seems to lack. It would reinforce the incalculable value of 'mothering' in the growth and development of children. Would paid household work also seem more valuable and attractive to men and boys? Sharing household tasks, while common to some degree in many cultures, is seldom so equitable that the woman does not spend many more hours involved in household work than does here mate.

Programmes which promote women's self-reliance and power rather than welfare and dependency may also produce stress within the traditional patterns and relationships of family and community life. Therefore, women's programmes must also recognize and address the problem of helping men and women cope with tensions that may arise with shifts in power and in roles. To do this, many women's programmes must include men, not to lead or dominate but to participate, debate and share in the development of decision-making mechanisms as equals.

(6) *Educational and extension activities for women must meet their needs for increased managerial, organizational, entrepreneurial and decision-making skills, along with technical skills related to food production, small-scale industries, etc.* According to FAO, the percentage of women in Africa among those taking part in informal instruction programmes was as high as 100 per cent in the field of home economics, and 90 per cent in nutrition. Outside these two areas, however, the percentages were far lower: 50 per cent of those being trained in the craft trades; 20 per cent in courses on livestock farming; 15 per cent in agriculture; and 10 per cent in management of cooperatives.

Although in most developing countries women account for the majority of the agriculture work force, they are still frequently ignored under training programmes. And in many countries, although it is women who dominate the informal sector (the small-scale trade in goods and services not usually counted in national economic statistics) they are seldom given any opportunity for formal training in marketing or management. They may also be denied access to credit, legal assistance, and land ownership.

(7) *Women's programmes must recognize and address health, nutrition and family planning issues in each particular country or community.* Society has traditionally given value to women's maternal role, yet this value is not always borne out by the support given to women to carry out this role. Many women still die in childbirth. Many give birth to low birthweight babies because of maternal malnutrition and a heavy workload during pregnancy. In many countries a lower value is still given to girls than to boys. In at least one country (reported by WHO) there is a significantly higher prevalence of malnutrition among female -children.

The special health needs of women are primarily related to their reproductive role. Maternal mortality accounts for the largest proportion (or nearly so) of deaths among women of reproductive age in most of the developing world; where this problem is most severe, many maternal deaths go unrecorded, or the cause of death is not specified. Only about one half of WHO's member countries can measure maternal mortality. Appropriate care during pregnancy and childbirth is vital to women's health and well-being. Failure to receive such care leads not only to death of the woman and/or her child, but also to debilitating conditions and infections which lower the quality of life of women. Malnutrition, including anaemia, is a serious health problem particularly of women who have too many closely spaced children. Related to maternal malnutrition and anaemia are the hazards of diseases and infections including intestinal parasites and malaria. Women engaged in heavy labour during pregnancy will have a mean weight gain of several kilograms less than other women with a similar food intake. Traditional birth practices can affect the health of mothers both positively and negatively. Uncontrolled fertility also aggravates many women's health problems. Abortions, both induced and spontaneous, are widespread, and illegal abortions kill up to 200 000 women a year.

In addition to their special health needs, women also carry the primary responsibility for health of their families and communities, both formally and informally. It is women who are expected to be health educators; to teach good health practices to the next generation; to create a home environment that supports health, from providing nutritious meals to clean water; to ensure that babies are immunized, taken to health care services when needed, and to care for any sick or elderly in the family. Women often serve as traditional birth attendants who deliver most of the babies in the developing world. They also provide most of the volunteer help in clinics, hospitals, and community organizations. Furthermore, in many places they are being recruited to provide primary health care to their community, in many instances with little or no financial reward.

Thus, the need for women to have access to health care and services, as well as educational opportunities in health and nutrition, is of great importance. No women's programming can afford to exclude this component. It has been shown repeatedly that the higher the level of a woman's education, the fewer children she is likely to have and the later she will start childbearing, while there is striking evidence that the woman's level of education is one of the most significant factors in the health of her children.

(8) *Programmes to promote rural industrial development to lessen the migration of men to cities must be planned with due attention to environmental concerns so that they do not destroy the area's ability to produce food, to provide water, fuel and a decent quality of life.* Unless this happens, jobs for men may only add to women's already difficult problems. And if women are to be employed in rural industries, they must not also be expected to carry out all the traditional household and food-producing tasks. Along with the industrial jobs must go labour-saving practices and devices for the home. Extension programmes should include ways to help families to look at the changing role of every member of the family from child to grandparent, including typical sex-related activities.

(9) *Extension programmes for rural youth of both sexes should include a strong emphasis on development education.* Such programmes can create an informed body of citizens for the country, and also promote knowledge and understanding of our global interdependency, environmental concerns, initiatives for world peace and equitable distribution of resources. Development education provides a worthwhile focus for 4-H and other rural youth activities especially during this International Year of Youth.

(10) *Existing educational structures and programmes must be used in ways that are effective and appropriate to meet the needs of women.* Formal and non-formal education programmes in home economics in many countries are not as useful as they should be in preparing women for the lives they will lead. Some school home economics programmes persist in educating students to prepare meals, which they might eat or serve at home, in kitchen workshops containing the kind of equipment they can never hope to own. Often, home economics is open only to girls although it is recognized that many of the skills included in the curriculum are equally important for boys. We need to question the validity and usefulness of such educational programmes. Are literacy programmes designed to meet the needs of women, particularly those confined to the domestic sphere who must be reached in ways other than by classes? Do existing agricultural programmes place equal emphasis on subsistence agriculture as on large-scale commercial farming? Are women trained to carry out extension and community education in areas other than food, health and nutrition?

(11) *Planning for women's programmes should include strategies for increasing integration and collaboration within self-help efforts such as primary health care, community education, industrial education.* Integration bodies should be established which look at needs of women without relying on the traditional lines of division of responsibilities according to ministries or departments. Efforts to do this in areas such as food and nutrition planning have not always met with great success. Nevertheless, where such integration can be achieved, without merely creating another level or layer of bureaucracy, agencies and departments may be able to respond quickly to local needs with increased resources. Networking is being stressed among national and community groups and agencies. International agencies are as guilty as local groups of neglecting to liaise and communicate with others involved in development work.

Cooperation and collaboration are facilitated when international and national systems of information storage and retrieval, and clearinghouses for materials and project information are strengthened. Can we design ways to provide materials, reports and information more quickly, widely and inexpensively than those which exist at present?

(12) *Women's programmes must reach the 'unreachables' and those most deprived of access to education and services through new media technologies as well as traditional media.* If we are to reach those most in need of programmes for an improved quality of life, we must not only make better use of new techniques, technologies and approaches (such as social marketing), but we must reinforce and encourage traditional communication forms such as song, folk drama and storytelling. If women still retain their primary socializing and educative role within the home and family structure, how can we provide support to them to carry it out?

For those committed to working with women in development there are challenges on every side. The Women's Decade has heightened awareness that problems and situations peculiar to women do exist. Governments and individuals have acknowledged this and committed themselves to action. It is our job to continue efforts which are effective, to continue to monitor and evaluate on-going activities, and to plan with enthusiasm and creativity the needed extension programmes for the next decade.

12

Extension Strategies Involving Local Groups and their Participation, and the Role of this Approach in Facilitating Local Development

JOHN F. A. RUSSELL

Agriculture and Rural Development Department, The World Bank, Jakarta, Indonesia

INTRODUCTION

Agricultural extension can be defined as the provision of knowledge and skills necessary for farmers to be able to adopt and apply more efficient crop and animal production methods to improve their productivity and living standards. The majority of conventional extension systems take new technologies developed at research stations and pass them in a top-down fashion to farmers following varying amounts of local testing that may or may not involve a farm trial. Much of this research is single-commodity oriented, and only with the advent of farm systems research over the last decade has the importance of looking at the whole farm situation been realized in designing extension programmes for small-scale, often largely subsistence farmers. Where production gains have been achieved by conventional systems, they have largely accrued to the larger farmers with more resources, who, with all the factors of production at their command, have been able to adopt the research recommendations with little risk. Thus, to ensure that the benefits of extension reach the mass of small farmers, it is essential that extension messages (and hence the generation of new technologies) are relevant to farmers' needs; this demands more active participation in the validation of new technologies by farmers themselves. It also means that much more attention has to be given to economic and socio-cultural factors than has been done in the past.

Furthermore, to service large numbers of small farmers effectively is costly — rarely can a country afford the luxury of having more than one village worker per 600–1000 farmers — and so to reach all farmers one has to work through a group approach. More importantly, however, is that for sustained development one has to foster the creation of active participation, and self-reliant farmer organizations. Governments and change agents have tried to take on far too much themselves, have often not recognized these problems of the small farmer, nor seen him or her as having any initiative (McNamara, 1983). The benefits of aid-funded donor projects have thus often been short-lived, as no self-reliant capacity resting on permanent indigenous institutions has been built up to sustain them, once the major investment period is past. Thus, not only are group approaches essential to reach the large mass of small farmers, but also to ensure that sustained local development that benefits the majority takes place. Once groups are formed the small farmers have a voice, which has to be taken notice of by governments and the agencies which serve them.

This paper draws on experience from a number of World Bank financed projects that illustrate some of the issues concerned with group development strategies for extension purposes.

APPROPRIATE METHODOLOGIES FOR FOSTERING GROUP FORMATION AND PARTICIPATION

One of the most successful projects that the Bank has been involved in lies in the Southern Region of Mali, where the *groupement villageoise* approach emphasizing farmer participation has been grafted on to a well run commodity-driven integrated approach to development. This is based on cotton, but now includes related food crops, and has been developed by the French Company for the Development of Textiles (CFDT) and is now run by a Malian Company (CMDT). Like the majority of successful interventions in rural development, the *groupement villageoise* approach has been developed and modified over the past fifteen years, and has been inspired by the work of Guy Belloncle (of Tours University, France), who has been personally involved with the approach over this period.

According to him: 'The precept of this new model of extension services is that African farmers, far from being locked into the use of out-of-date techniques and resistant to any innovation (as some would have it), are indeed responsible adults, aware that they must alter their traditional farming practices and therefore be prepared to learn new ones, *provided that they understand where they are being taken*. There is no shortage of examples showing that when technically feasible, sociologically acceptable and economically profitable innovations are proposed to them, the new practices are quickly adopted. Hence farmers cannot continue to be blamed for something that is not their fault. The approach must be changed and a *true dialogue* be held, not with individuals selected from outside but with *existing communities* (that is, villages and districts), to propose a type of rural development that will ensure the survival of the group and leave no one behind.' (Belloncle, 1985, page 6.)

Belloncle has also shown how 'group instruction' should take place in three phases: self-analysis, self-programming and self-evaluation. This is carried out by the farmers themselves prompted by iterative discussions with extension agents and researchers on their problems and the options for dealing with them. In the Mali case, self-analysis comprised seven evenings of listening to how the farmers themselves analysed their agricultural situation and how, with the agent's advice, the farmers might improve it in each of several villages. Belloncle comments on the extraordinary clarity with which farmers saw their main problems in the village, and suggests how to draw a representative village sample for conducting such analysis in preparing a project, and that at least as much weight be given to farmers' perceptions and expectations in project design as that of outside experts. Although this iterative process is time consuming, it lays a much surer foundation for development, and the bottleneck is not time but the ability of extension workers to perform this task, and their need for appropriate training in analysis, awareness, sociology and pedagogy.

In the phase of self-programming, the technicians have first to respond with solutions to the problems raised in a consistent overall strategy. The villagers then decide on who will make the first test of new techniques recommended, and in what order they will do them. The key here is that the villagers select their representatives to carry out the trials, and so will feel confident in reviewing their experience, rather than extension staff selecting key innovators or pilot contact farmers who the villagers may not feel they can emulate. There is also often a sequential combination of improved practices to be adopted, rather than simply individual impact points. Mann (1978), discussing the adoption of wheat technology in Turkey, has shown that disaggregating a research package into clusters of practices adopted in an agronomically and economically logical sequence relates to typical farmer behaviour and provides a guide for both experimentation and extension recommendations. We should neither have a rigid package that will only be wholly adopted by a few farmers, nor isolated individual impact points. The appropriate sequence has to developed in discussion with farmers themselves, and varied for different types of farmer. The designation of literate farmers helps in this selection, as they can more easily interact with the technicians and record quantitative data on which the financial calculations on returns can be made verifiable by all. There may be a trade-off here as often the literate farmers are the larger-scale ones, which points to the need to target functional literacy campaigns to ensure all resource groups of farmers are included.

The self-evaluation phase can be defined as the time when the farmers' representatives report and discuss with their group and the technicians their experience, which all have been reviewing as it went along. If the new techniques are more effective, the objective obstacles to their being more widely adopted can be concretely discussed. This self-evaluation phase leads to further self-programming.

This can be a viable methodology for ensuring stronger farmer participation, and ensuring recommendations are relevant to their needs. Much, of course, depends on the cohesiveness of village society, which was generally strong in the Mali case. Where village societies are divided, one can only work with the differing subgroups. The programme in Mali was aided by functional numeracy and literacy campaigns, which were well received once a defined need for them existed. This was not only for facilitating adaptive trials work but more for the overall benefit of farmer associations that were formed in each village with responsibility for the final stages of input supply, savings and initial marketing functions. The latter provided a margin to the groups on each kilogram of cotton that provided the 'engine' for the system to take off, and spawned other internally sponsored development activities.

A parallel methodology to the one used for the *groupement villageoise* approach, drawn up on similar lines, has been the one used in Mexico's PIDER (Integrated Progam for Rural Development) (Cernea, 1983). Experience in PIDER has been mixed, but it illustrates the need for competent extension technicians and sound technical advice that has to be available (and wasn't always) for the interventions under this programme. Successful group activities in extension were developed under Mexico's Plan Puebla, which has been drawn on in developing group extension in Mexico's PRODERITH program for Tropical Agricultural Development. In the first phase of this programme some 946 groups were developed, and in addition to extension activities they have gradually taken over the operation and maintenance of rural works in their areas.

BUILDING ON EXISTING INDIGENOUS INSTITUTIONS FOR GROUP FORMATION

Many of the more successful extension interventions involving group formation have been drawn on existing traditional groups, which have too often been ignored in donor-funded development programmes. A good example is provided by Indonesia, which has been introducing the Training and Visit (T & V) system, and has been able to build on a number of such traditional groups. Rural people in Indonesia have lived and worked for centuries with strong communal groups or cooperatives such as *subak* (farmers' associations formed for use of irrigation water), *gotong royong* (farmers' self-help association), and *mapulus* (farmers' mutual assistance association). These have a clear membership, and high loyalty to the leader, and for introducing T & V extension, these farmer groups have become the focus for guidance and technology transfer.

Historically they had been rather hierarchical with a tendency to operate in a top-down fashion, and so when the T & V system was introduced the traditional leader often became the contact farmer. Now, however, they have moved to identifying sub-groups in adjacent field areas and matching up farmers with different resource levels to have more representative contact farmers. They discuss and agree their seasonal extension programme, decide on technology appropriate to their situation, and monitor feedback from questions posed at the regular visit of the extension worker. In this way they have become much more participatory. The village worker receives special training in maintaining group dynamics, and the local administration gives active encouragement (Sukaryo, 1983).

When new groups are formed, they receive the tacit agreement of the local power structure, which is an important concern when initiating new groups. This need to gain the support of local leadership without directly involving them in running groups is a key one. Coletta (1979), writing about popular participation through the Sarvodaya movement in Sri Lanka, distinguishes the new informal leadership of such development groups as the institutionalization of a dual leadership system: one of symbolic governance and authority through the traditional formal leadership that is non-participatory, and another of mass action through informal groups and leadership under the approving eyes of that same symbolic structure. This arrangement of dualism has been viewed as a strategic stage in moving decision-making power from a handful of village elites to more popular mass participation. However, it highlights the importance of getting the support of both traditional leadership and the local administration, while retaining powers of action within groups themselves, which need to be small enough to be mutually cohesive and to accept collective responsibility.

The final example, which is again based on an adaptation of the T & V approach, is from a pilot project in the Midlands of Zimbabwe, that has been running for the last two years. Here again there is a strong tradition of work associations for sharing agricultural labour, and more recently of savings groups, that have been especially strong among women. Here, farmer groups organized for extension purposes are normally sub-divided when numbers exceed thirty, to keep practical demonstrations effective and retain the groups' cohesiveness. A key feature is that the final stage of dissemination of improved technology is done by farmer representatives themselves, two of whom are selected to attend each training session fortnightly and to report back to the rest of the group. This is a tenet of the T & V system but rarely is it institutionalized and done so effectively. The chosen contact farmers rotate amongst group members depending on their individual attributes related to the topic under discussion. This enables several members of the group to be both trainee and trainer, and by going together it reinforces their mutual knowledge of the lesson being discussed. The fortnightly training sessions, which other farmers may attend as well as group leaders, are conducted on members' plots, and group leaders rotate their training of fellow group members around different member farms. Slogans in the vernacular and songs about the lessons have all helped to motivate keen participation by all members, especially the women. Whereas less than 10 per cent of farmers were in groups before the pilot scheme started two years ago, about 70 per cent were covered by groups by mid-1985.

ISSUES IN THE SELECTION OF GROUPS AND THEIR LEADERS

The main issue in group formation is that the group has to recognize the need and the benefits to be obtained from forming a group, and in maintaining it. This means care has to be taken in forming the first groups, and in demonstrating to others the value of having them. Group formation which is politically or administratively imposed from above rarely works, and the groups will be shortlived if members do not derive benefit from them. There has to be relevant technology that group farmers can adopt to their advantage to ensure active participation; and this has to be revised and enhanced as the group develops. Relevant interventions are thus needed for initially starting groups, that will strengthen as they realize the benefits of a new relationship and improved interaction with research and extension staff, as well as starting to have more of a say in their own destiny which governments have to listen to.

Likewise, farmers need to be stratified according to their resources, so that technologies extended are relevant to their conditions. Hence, the value of dividing farmers into recommendation domains as has been done by Collinson (1984) in his farm systems work in East Africa. Such groups can be sub-groups of a larger group, their nature varying according to the social structure and the cohesiveness of the society. The Zimbabwe example demonstrated the need to split groups up when numbers exceeded thirty for effective extension communication, but the size of the group will vary according to traditional social factors, such as the size of family or kinship groups. In Mali, it was necessary to have separate women's sub-groups due to their place in indigenous society and their varying responsibilities in labour and crop division in the farming calendar, whereas in Zimbabwe mixed groups were found to work very well.

We have already discussed the value of building on existing groups, but this requires adaptation, since an existing group formed for a different function may not provide a suitable structure for extension groups. A strategy involving sub-groups may be pertinent here, but it is necessary that the functions and role of the sub-groups are fully recognized by the leadership and members of the larger existing group.

Members must take part in leadership selection, and rotating leadership as used in the Zimbabwe groups has much to commend it, especially when associated with their acknowledged varying technical expertise in different farm enterprises under discussion.

TRAINING, COST EFFECTIVENESS AND SUSTAINABILITY

Experience shows that extension workers need special training in group activities. I consider this to be an important, albeit often neglected part of their role, though others have suggested that different people specially trained for the task should fulfill this role. However, this creates the need for having good trainers in this sphere. Further, all the examples cited have mentioned the importance of providing strong technical recommendations well adapted to local socio-economic conditions. This also calls for better trained staff, and the need to constantly upgrade them, so that they can diagnose and analyse with farmers their problems, and present options (not just one recommendation) for their solution.

Turning to the question of training farmers, it is clear that this is best done practically in the farmers' fields. It is also apparent from the Zimbabwe experience that farmers themselves can be effectively trained as selected group representatives to pass on new technologies to their fellow farmers, and in this they can often be more effective than the village worker himself. The Mali experience showed the need for and value of adult literacy and numeracy campaigns; and it is interesting that earlier attempts at this done in isolation with no specific goal in mind had failed, but that when a real use and purpose for the literacy in the *groupement villageoise* approach was recognized, the demand and recognition of the value of such training mushroomed.

Effective development of groups does mean that fewer face-to-face extension workers are needed, especially once the system is running. In Zimbabwe, one worker now handles 800–1000 farm families, but in other places ratios of 1:200 or 300 still exist which governments cannot afford to replicate or even maintain. In the Second Phase Mali Sud Project, where the *Association Villageoise* are fully established, instead of having one extension worker to each as before, the numbers are more than halved to three workers for seven *AVs*. This move both reduces the recurrent cost burden on government, and also develops institutions at the grassroots level that are more self-reliant and hence better ensures the sustainability of the new approach and the investments made. This is a typical evolutionary step in the development of extension services that need less face-to-face contact as farmers become more sophisticated and can make better use of the media. Active group development, however, helps to spur on this process.

In a number of cases, governments have tried to involve farmers as paraprofessionals in helping to disseminate new technologies, and paid them a small stipend to give up part of their time to do this (Esman, 1983). In a recent instance of this at Wedza in Zimbabwe, where a three-year pilot programme was drawing to a close, however, it was evident that once government started paying the selected farmer, he became more identified with government than his group, and more concerned with improving his terms of service than the well-being of the groups he served. Contrasting this with farmer group representatives who are not paid anything by government, on the T & V pilot programme in the Midlands of Zimbabwe, it is clear that the latter system is far preferable, and the farmers involved are proud to remain fully identified with their groups. If there is a need to cover their expenses, this should be done by the group, not by government. An interesting example from Ethiopia is that Peasant Association members help to cultivate the fields of their farmer representative, to make up for some of the time he is away acting as their representative.

It is interesting to note that the successful groups once established have developed into stronger more permanent organizations that become involved in more facets of local development. The *AVs* in Mali were set up to handle part of the input supply and marketing functions for which they obtained a payment since they reduced the costs to government and were given a share of the margin for their efforts. This proved the engine, on which various productive and social developments took place according to the villagers' own priorities. The importance of numeracy here assists democractic control, because there have been instances of abuse of their position or misappropriation of funds by leaders. The groups in Indonesia, likewise, pass through a process by which they evolve, moving from elementary, through advanced to *madya* groups. The latter status is reached when the majority of farmers have adopted the recommended improved practices. They then move on to a cooperative system, known as *utama*, which has a strong legal foundation and more formal constitution, and is a logical evolution of the groups moving from an essentially subsistence to a cash economy.

CONCLUSIONS

In summary, extension strategies need to involve groups not only to ensure more farmers are reached at lower cost, but more importantly to provide viable local institutions that can help farmers to be more self-reliant, and to sustain development investments. The way groups develop will be conditioned by existing social structure, and the nature of the environment. The role of extension staff in fostering this is an important one, for which special awareness and training are needed. Existing institutions have too often been neglected in rural development, but they may need careful adaptation to fulfill an expected new role in the extension field, possibly requiring sub-groups, since for extension purposes they should not be too large. Successful group development has a major role to play in both sustaining development at the grassroots level, as well as reducing costs to government, which is less well equipped in any event to play such a role permanently. Sound technology, well adapted to local conditions, is a necessary prerequisite to build on, requiring better trained extension staff, better attuned to farmers' needs and prepared to work as equals *with* farmers rather than *to* farmers (Russell, 1985). For we have to remember that all research and extension activities are only a means to an end: 'increased productivity on the farm, and the well being of the farmer'. The farmer is central to all these activities, and to disregard his or her vital input is to guarantee that much research and extension activity will be irrelevant and not a cost-effective use of scarce budgetary resources.

13

People, Power and a New Role for Agricultural Extension: Issues and Options Involving Local Participation and Groups

MALCOLM J. ODELL JR

Synergy International, Amesbury, Massachusetts, USA

'The extremes of rural poverty in the Third World are an outrage' (Chambers, 1983). Progress in agricultural extension seems tediously slow. A crop survey in Botswana in 1948 described agricultural conditions and proposals for change that differ insignificantly from those of today (Parish, 1948). Some even argue that development assistance has done more harm than good, that 'hundreds of millions of desperately poor people throughout the world have been hurt rather than helped by economic development', that economic aid, as presently administered, can be an obstacle to development, and that worldwide almost none of the 'small farmer credit' programmes reach the small farmer (Illich, 1969; Adelman and Morris, 1973, page 192; Lappé *et al.*, 1980, page 57).

THE NEED FOR NEW STRATEGIES

In recent years, re-examination of past failures and the intransigent nature of poverty in the Third World have led to calls for new approaches. Proposals often cluster around two major themes: the decentralization of decision-making power to local institutions; and the stimulation of private enterprise/entrepreneurship. Many (e.g. Korten and Young, 1978; Korten, 1980; Moris, 1981; Korten, 1982; Rondinelli, 1982; Chambers, 1983; Esman and Uphoff, 1984; Uphoff, 1984) suggest shifts away from large, capital intensive projects. Their research suggests, instead, the promotion of small, locally controlled micro-projects and iterative, trial-and-error processes which allow local producers and institutions to learn from experience.

On the private enterprise front, parallel proposals call for emphasizing micro-enterprises, providing very small loans to large numbers of borrowers (Smith and Tippett, 1982). Whether community- or individually-oriented, proposals show a remarkable similarity in their shift of emphasis to the most localized level, and the scaling down of projects to a level far below those traditionally promoted by major aid donors and national governments.

WHY AN EXTENSION EMPHASIS ON GROUP APPROACHES?

It may be asked why agricultural extension should place particular emphasis on group approaches when there are good arguments for privatization and developing the role of the individual entrepreneur or small farmer. The emerging data suggest that group approaches are particularly noted for:

(a) Efficiency: in terms of the resources required to reach large numbers of beneficiaries, group approaches offer widespread coverage and reinforcement at low cost;

(b) Effectiveness: group approaches result in better projects for more people in less time than do individual approaches, particularly when built on existing institutional or cultural foundations; and

(c) Equity: in protecting the resources, rights, and economic security of the poorest members of society, group approaches ameliorate the negative effects of privatization.

Few will argue with the contention that group approaches are efficient in reaching large numbers of rural people at low per capita cost. A sizeable school, however, including Hardin and other proponents of individualism and a free market economy, claims that group approaches particularly as they affect the management of the commons, are anathema to long-term development. They point to failures of group initiatives in community development, cooperatives, water management and group ranching around the world. Others claim that by involving more people in decisions, valuable time is lost and important technical considerations are jeopardized (Hardin, 1968; Hardin and Baden, 1977; Oxby, 1982, 1983).

Those of the 'tragedy of the commons' school can legitimately point to failures. What is lacking in the debate, however, is a thorough examination of why and how things have gone wrong and what can be learned to overcome the problems identified. Essentially, do projects fail because, as Hardin suggests, the system of collective management is inherently flawed and doomed to failure? Or do they fail because of external factors which are predictable and controllable? If inherently flawed, then the survival of viable community management systems over centuries is, indeed, an anomaly.

The brevity of this paper allows only a cursory response to the underlying issues raised by these arguments. Here we will examine briefly: some successes in community and group management; some lessons from experience; and, equity and the distribution of power.

Some Successes in Community and Group Management
Local government is alive and well around the world. For centuries, villages have done a remarkable job in providing for the common good by allowing individual democratic participation in decision-making and economic security through the utilization of complex individual, family, and community-based production and distribution systems. Local institutions across the globe demonstrate that a sound traditional base exists for the sustainable and equitable management of village commons. Their record in community management, and their potential for management of such communal resources as grazing and forest reserves, is well established. Examples are readily available from every continent — including the *souming* system in the crofts of the Hebridean Islands in NW Scotland; pasture management in Peru; the *maboella* system of Lesotho; the *dinaga* and the *ohambo* systems among several different tribes of Namibia and Botswana; Sherpa *naua* and *shingo naua* systems of village and forest guardianship; *panchayat* management of village forest lands in India; and many more (Ashok, 1978; Hitchcock, 1980; Odell, 1980, 1982; Odell and Odell, 1980; Gilles and Jamtgaard, 1981; Devitt, 1982; Runge, 1983; Uphoff, 1984).

Extensive data exist to respond to the criticisms of those who claim that participation and community involvement in decision-making delays and politicizes sound development planning. Over 150 case studies examined by Cornell University, for example, demonstrate qualitatively and quantitatively that meaningful community involvement produces better projects reaching more beneficiaries within less time than do those without such participation (Eberts and Kelly, 1983;

Esman and Uphoff, 1984). Furthermore, Runge (1983) demonstrates through economic modelling, basic fallacies in the 'tragedy of the commons' concept and a rationale for successful communal management.

Some Lessons from Experience

Botswana provides a case in point. In this Southern African country a group ranching programme initiated by the government and USAID sought to 'establish socially acceptable and economically viable groups of smallholders to engage in improved agricultural production'. The project brought in technical range and livestock management specialists to organize some 16 group ranches. After three years it was deemed a total failure.

The Government of Botswana, drawing heavily on rural sociology and development anthropology studies, asked AID for a new project — with an identical purpose — which called, instead, for group development, community organization, and applied social science specialists. Key elements of the strategy were that:

(a) the extension service was given overall responsibility for the group development effort, with the project's expatriate staff serving in training and consulting roles to build community organizing expertise among field staff, and local agricultural staff providing technical backstopping;

(b) simplified social science techniques were used to identify existing community and farmers' groups (based on traditional social relationships or common interests) rather than to organize new groups; and

(c) financial resources were provided in the form of matching mini-grants for community-initiated projects.

The result was that group and community projects multiplied faster than government could monitor. Within three years over 1200 group projects were identified and the number increased monthly. One community that had sent away the first project team, claiming that the group ranch proposed was not suitable for it, had completed[12] agriculture projects and was planning a group ranch of its own innovative design (Odell, 1980; Odell and Odell, 1980; Willett, 1981).

Similar examples of successful extension initiatives building on local groups and traditional institutions can be cited in Nepal, where a traditional forest guard system has been re-introduced to halt hillside degradation, and the Philippines, where the National Irrigation Authority has provided community organizers to strengthen local farmers' organizations and help them take charge of local irrigation efforts (USAID, 1978; Korten, 1982; Odell, 1982).

What are some of the determinants of successful community and group management systems? Some that emerge from current literature and field experience such as that described above are:

(1) Traditional, locally understood, and locally managed structures (rather than new institutions).
(2) Adaptation rather than replacement of traditional systems when change is essential.
(3) Local decision-making authority on priorities, methods, day-to-day operations, and the adaptation of existing systems (rather than technically 'correct' solutions based on outside assessments).
(4) Relatively small cohesive community, neighbourhood, or users' groups based around a regularly used common local resource.
(5) Local, decentralized revenue and/or tax generating authority.
(6) Loans or matching grants to augment locally generated funds (preferable to outright grants).
(7) Clearly defined, adjudicated community/resource boundaries.
(8) Legal authority of communities over resources.
(9) Land-use planning directly involving the communities, supported by professional expertise.
(10) Institutional development and training support to strengthen local institutions.

Equity and the Distribution of Power, or 'The Helping Hand Strikes Again'

Have the results of development through individual approaches and privatization been as sanguine as economic theory predicts? Certainly many private owners have made great successes of their holdings and are today managing them in an economically productive and ecologically sound manner. Other evidence concerning privatization in agriculture and range management is less than encouraging. Numerous private ranches in Botswana and Kenya, for example, are showing extensive degradation. The transfer of communal land rights to individual and commercial interests across Africa has exacerbated rural landlessness, poverty, and unemployment (Odell, 1980, 1982; Odell and Odell, 1980; Sandford, 1980; Devitt, 1982). 'There is no surer way of depriving a peasant of his land than to give him a title to it which is as freely negotiable as a bank note' (Simpson, 1967).

One historian and philosopher on the dynamics of modernization has pointed out that '... premodern societies are much more diverse in their institutions than modernized societies ... [T]hey can work together in teams with mutual respect ... [which is] in many ways well suited to modern life' (Black, 1985). When we undermine people's incentives to work collectively, we undermine their ability to increase their power to feed their families and lead lives of productivity and dignity.

The basic issue, then, is one of social and economic equity for smallholders, and the bottom line is power. If we believe that continuing poverty and disparity in the Third World is an outrage, then we must take the necessary steps to bring about the kinds of reversals in power relationships that are required. The assistance currently provided is not bringing about the change that our rhetoric proclaims. In the words of Ivan Illich, 'the ploughs of the rich can do as much harm as their swords' (Illich, 1969, page 1; Chambers, 1983).

EXTENSION STRATEGIES INVOLVING LOCAL PARTICIPATION AND GROUPS

What can we do? How can we address this issue of power within the limitations and constraints of our profession? As noted above, the governments of the Philippines, Nepal, and Botswana have, in their own ways, attempted to answer these questions by instituting extension-based programmes to help local communities and groups to organize, increase their power, and take charge of their economic lives.

Extension professionals (as we look to the year 2000) are in a unique position to contribute substantially to major changes in rural development practice. We have access to hundreds of thousands of trained workers located in countless villages around the globe. Their considerable expertise lacks only the direction and tools to address the organizational and power issues of the rural areas — issues tied closely to decentralization, local participation, and group approaches.

The body of knowledge necessary to strengthen and support group and community-based action is far removed from the skills that we have generally imparted to agricultural extension personnel in the past. It will be essential to increase inputs from the applied social sciences —

especially rural sociology and development anthropology. In addition, we have another resource readily available: an under-utilized and neglected tradition of community development. Now is the time to examine the implications of weaving together these two professions.

Community development during the 1960s and 1970s became bogged down in the provision of non-productive, non-economic social services that did little to address the fundamental issues of food and income. Its expertise in the critical analytical and organizational skills associated with bringing people together to solve their common problems has yet to be meaningfully tapped for economically productive programmes. The movement is, some say, in need of new direction.

Conversely, agricultural extension has focused almost exclusively on the economic and technical issues of income and food production, frequently working exclusively with the individual farmer. Expertise in analysing existing social structures and facilitating group and community action to achieve economic ends is minimal at best. The profession has many of the technical answers that farmers and farming communities need, but is not well equipped with 'people skills' and organizational know-how, particularly in facilitating and supporting effective and successful grassroots community and group initiatives.

What are some of the specific strategies and tactics that such a merger might employ?

Grassroots Farming Systems Research and Extension for the Landless and Near-Landless

Extension professionals have always sought to reach the 'grassroots', the small farmer. Success in this area has been limited, however, and relatively wealthy farmers have been the first, and sometimes only, beneficiaries reached. The time has come for extension to use stream-lined farming systems research and extension methods, to identify clearly the problems, priorities, resources, and institutional/social structures within which the poorest groups carry out their quest for survival. Such methods draw upon the applied social sciences and the rapid rural appraisal techniques popularized by Chambers that have now been widely tested in the field. Staff training manuals and materials to support these methods have been prepared for replication globally. It is time farming systems research moved out of the test phase and into the field, using the powerful extension network and focusing on those we have been least successful in reaching in the past (Chambers, 1981; Shaner et al., 1982; FSSP, 1985).

Institutional Development to Strengthen Indigenous/Traditional Institutions

A second strategy worth examination by the extension profession is, through collaboration with experts in community organization and social change, to work with host governments to strengthen and support local indigenous institutions. Extension personnel have always known that they will never succeed without the cooperation of the local chief, headman, or council of elders. What they have not yet done is to work to strengthen such critical institutions, rather than simply exploit them for their own purposes. Some specific tactics which might be examined include ensuring and/or strengthening the legal authority of local communities over their resources, including clearly defined, adjudicated community/resource boundaries; the legal authority of communities to levy taxes or fees on their members; and land use planning and management skills for communities.

A Learning Process Approach Embracing Error

Extension professionals have already learned that 'top-down' planning does not work. What is not so clear is that the profession can move away from a 'prescriptive' approach that demands professional planning to an 'adaptive' approach that embraces error and helps communities learn from their own mistakes. It is time to examine the implications of moving away from large-scale, 'integrated rural development' strategies that over-professionalize development and depend on limited numbers of 'mega-projects' that are beyond the management capacity of local institutions. 'Micro-projects' that are easily implemented locally are worth development and multiplication. From these, communities can learn the skills necessary to make each succeeding project work better than the previous one (Korten, 1980; Esman and Uphoff, 1984).

Extension as a Facilitator in Empowering the Powerless

If we are to have any impact on redistributing power in the Third World, as the profession with the most direct access to the rural poor of perhaps any other, we must consider how extension personnel can serve a facilitating function in empowering the powerless. This must be done, not to bring the poor into the development process, but to ensure that the poor achieve the power they need to direct the development process in their own interests (Lappé et al., 1980).

What is needed is a process that brings more of the insights and skills long available in the worlds of the applied social sciences and community development to the agricultural extension community. Some of the specific skills that might be imparted include: facilitation and group process skills for supporting groups, rather than directing or manipulating them; management, goal-setting, decision-making, negotiation, and meeting skills for members; and procedures for maximizing membership participation.

To embark on such a strategy is not a radical process, but an organizational one based on efficiency, effectiveness, and equity. We would be doing no less than expanding fundamental processes of democratic decision-making and pluralism that we take for granted — and demand — in most other contexts. To do otherwise is to invite rather than discourage radicalism in the Third World. A constructive extension process designed to confront the issue of power directly and help communities and farmers' groups around the world to take charge of their own development is a healthy antidote to the pervasive prison of poverty and the frustrations, anger, violence, and despair that it breeds.

Discussion

The strategies which should underlie extension work were discussed in relation to the three areas covered by the papers: the transfer of technology, extension work among women, and with local groups. In each case, it was recognized that there would be a need for a set of strategies. In detail, these would have to be appropriate to particular situations, their needs and resources, differing national policy objectives, and varying organizational philosophies. Even so, at the level of basic principles there could be much in common between different countries, whether developed or less developed, although the focus of the discussions was primarily on the latter. In all cases, the strategies would need to be congruent with social differentiations within rural societies and the consequent implications for the organization of extension services, especially the training of extension workers.

In discussing the transfer of technology, there was much agreement that the prime function of extension is the transmission of relevant knowledge and information to rural people; the addition of other functions inevitably reduces the efficiency of an extension service. However, if other essential elements (indicated in the paper by Prof. Bunting) are missing, little benefit is likely to occur from knowledge transfer alone. Thus, at least in the short-term, extension services would have to assist in providing these. Their aim, however, should be to shed the other functions by keeping their government authorities aware of the problems which they cause, and persuading governments to adopt policies which encourage the development of other public and private organizations to provide the particular services in appropriate ways.

Two related dichotomies formed the basis for much of the subsequent

179

discussion — the distinction, first, between commodity production systems (commercial crops) and food or subsistence production, and secondly, between resource-rich and resource-poor farmers. Relatively simple information and instruction, if allied to the provision of necessary inputs, credit, and an assured market offering fair prices, is generally sufficient for commodity producers to be efficient. However, subsistence farming systems are often much more complex. Here, farmers require extension work with a more educational approach which can lead not only to a knowledge of the technology proposed but also to an understanding of it and its underlying principles so that they can be enabled to make their own farm management decisions, now and in the future. The farmers who are the relatively resource-rich are likely to seek for any information they require, so that the obvious strategy is to concentrate extension work on the resource-poor. However, extension services should aim to serve both categories, for two main reasons: (a) it is largely from the resource-rich that rapid increases in production which enter the market are likely to be obtained; and (b) if the resource-rich are neglected by extension services in favour of the poor, the result could be socially disruptive, with the richer, higher class farmers becoming antagonistic to the advantages being gained by the resource-poor and to the narrowing of the disparities between them. One feasible strategy would be to organize separate extension services for each category — a wholly government-(public-) funded service for the resource-poor, and a service for the resource-rich farmers funded at least in part by themselves.

A somewhat comparable line of reasoning developed in discussing the relationships between extension work and women. It was recognized that low-cost technology, appropriate to local needs and knowledge, which could relieve women of the drudgery of much of their domestic and other work, is often lacking, or at least is not known among rural women. There is a need to disseminate what is known more effectively. However, this requires a better knowledge of women's needs and of their activities and allocation of time. It also requires a change in attitudes at government levels to ensure that interventions with a social value are perceived as of equivalent importance in rural development as those which lead to economic benefit. It is also important to recognize and manage the confrontation between men and women which can be created by extension work among women. The aim should be to increase and improve the activities of extension services without creating male hostility or antagonisms to the work due to the men feeling threatened by the prospect of losses to their social status and power. The required strategy should thus be socially acceptable, seen to be beneficial by men as well as women, and so coordinated that the men feel involved and motivated to support the extension work among women. In this, although some tension is possibly inevitable, one need is to organize problem-solving groups in which men and women can become increasingly conscious of their social roles in the development of rural communities.

This approach was developed further in the discussion of extension strategies involving groups and local participation. The use of groups in extension work is complex. To be successful, and if groups are to be able to sustain themselves, it is important that they should be integrated into the larger institutional frameworks.

In the three strategies discussed, there are several implications for the organization of extension services and their work. If their activities are concentrated on some members of a local social system (e.g., the resource-rich, or the men), extension workers may exacerbate tensions and inequalities in a rural community. The need, however difficult, is to involve everyone. To do this, more women should be employed in extension services. It may also be necessary for some extension workers, especially at regional level, to be given new specialist responsibility (e.g., regional group development officers).

One major implication is the need to improve the training given to extension staff. To work with resource-poor farmers is not only time-consuming but requires special skills. Since in many developing countries the availability of extension workers is limited, this might appear to be an unrealistic aim. However, in the long-run, knowledge-able and self-activating farmers can only be developed by improving their understanding of relevant technological innovations.

To increase extension work among women requires greater oppor-tunities for women to enter extension agencies. This can only arise if decision makers can be sensitized and change their attitudes regarding the capabilities of women and their willingness to employ them. It also requires raising the self-consciousness of women and changing their perceptions of their roles and their future opportunities — processes which need to begin while girls are still at school. In developing group work, the organizing skills of extension workers have to be developed and, since all groups are to a degree heterogeneous, the information presented has to forgo specificity in favour of more ambiguity. The training curriculum of extension workers of both sexes should also cross traditional boundaries (e.g., nutrition should be taught to men as well as women). Further, to develop and work with groups, and also with farmers conceived of in various categories, extension workers require training in aspects of applied social sciences. This is essential, for example, to enable them to handle the development and dynamics of groups, and to increase their awareness of the socio-economic details of a local situation by having the necessary skills to analyse a rural community.

Each of the strategies discussed implies a broader range of extension work among a wider spectrum of the farm and rural populations, which requires more and better trained extension workers. This has clear implications for the funding of extension work. It was agreed that farmers already pay, indirectly, at least for part of the extension services through taxes levied on export crops and the maintenance of low food prices. Farmers also pay (or would pay) for certain specific services (e.g., markets, storage, processing) provided by public or private sector agencies if they are convinced they can benefit from them, and this could lead to the improvement of such services so long as other essential services are available. Some maintained that direct payment by farmers for their extension services would not only reduce government investment in them but lead to improvements in their effectiveness. For example, in one of Africa's poorest countries, Mauritania, farmers have recently been required to contribute the equivalent of £1 per year towards their extension service. This has led to a dramatic increase in their expectations and demands upon the extension workers resulting in a service which is more clearly focused and client-oriented with increased benefits and reduced government expenditure.

However, payment for services and the encouragement of private sector agencies leads to problems of equity. This has to be considered in designing strategies. Private sector agencies and paid services are likely to support the development of cash crops, while public sector extension work would probably still be expected to assist the development of subsistence farmers and food crops, and to be concerned with the broader, more humanistic objectives of rural development. This could lead, however, to the existence of a dual extension system in which first-rate services are available, for payment, to some, while the remainder are only offered poor, second-rate services.

EXTENSION METHODS

14

Mass Media and Communications Technology

CHRIS GARFORTH

Agricultural Extension and Rural Development Centre, University of Reading, UK

CONTEXT

A Developing Technology

If information is a key resource in rural development, and the communication of information a major function of extension, it is no surprise that extension practitioners are looking with interest at the potential contribution to their work of mass media and communications technology. The speed and efficiency with which information can be transmitted by electronic means from one place to another, and the capacity of computer-based information systems to store data for rapid retrieval, continue to develop apace at a time when extension organizations are looking for more cost-effective ways of making useful, relevant information available to rural people. At the same time, there is a growing recognition of the need to support extension workers in the field, both by contributing to the general information environment within which they work, and, more directly, by providing them with extension aids and updating their own technical expertise and knowledge.

Developments in communications technology also increase the potential for rural people and communities to communicate with each other and to make their views known to government and commercial institutions. Information from the field can also be collected and analysed more readily at the local level, which facilitates current attempts to decentralize extension planning and evaluation.

Mass media and extension aids, ranging in technological complexity from film to flannelgraph, have been used in extension for many years.

There are three technological trends, however, which are significantly increasing their potential role. The first is the *decreasing cost, increasing reliability* and relative *ease of use* of much recent communications technology, all of which make the technology increasingly accessible. The development of computers (especially the range of increasingly powerful, compact and relatively inexpensive microcomputers) and video are two cases in point. Second, recent years have seen considerable technical developments in the *transmission* of information, both globally through a rapidly growing network of communication satellites which has the potential for linking people in virtually any two places in the world in a face to face dialogue, and more locally through cable networks and small-scale transmitters which open up opportunities for community-based radio and television services. The third trend is the *integration* of the various components of communications technology, especially the bringing together of the data processing and storage capacity of computers with the ability of telecommunication systems to transmit electronically-coded data extremely quickly.

A few examples will indicate the range of possibilities opened up by these developments:

(1) Field extension workers in remote areas can be linked with expertise and sources of information in central locations. Two-way radio, via satellite, is already well tried in the health field, enabling diagnoses and advice on treatment to be given by a doctor or midwife hundreds of miles away (Laflin, 1982). The expansion of telephone networks into rural areas offers similar opportunities.

(2) Research workers can carry out on-line literature searches on bibliographic data bases held in another continent.

(3) Viewdata systems link domestic television sets, through telephone lines, to information stored on centrally located computers. In the UK, a public viewdata system (PRESTEL) enables a user to call up the information required for display on a television screen, the information itself being regularly updated by various information providers. The ADAS is already using the system to make a wide range of information — agricultural news, technical advice, market prices, pest and disease forecasts — available to farmers (Houseman, 1981).

(4) Video is being used in a number of ways: making field recordings for incorporation into broadcast television programmes; bringing

technical information to rural families on mobile units or at rural training centres; in motivational work in community development programmes; by community groups in campaigning or lobbying activities; and in training programmes, to give instant feedback to extension trainees on their performance in practical exercises (Fraser, 1980; Berrigan, 1981; Bright, 1981; IIC, 1983; China and Langmead, 1985).

(5) Community television sets can be linked, as in India's SITE experiment and later developments, via satellites and land-based relay stations, to a central transmitter, enabling a mass audience of rural families to see and discuss centrally produced educational broadcasts (Karnik, 1981).

(6) Microcomputers are being used within pest and disease surveillance systems in South East Asia to localize the analysis of surveillance and agro-economic survey data. The possibility exists for linking these microcomputers, via the national telephone network, to a central mainframe computer where data could be compared and aggregated for further analysis.

The Institutional Context

The above examples demonstrate that communications technology offers increasing scope for localizing the production and distribution of media, for two-way interactive communication, for communication within and between rural communities, and for a more active role to be played in communication processes by people who have hitherto been regarded (in communication theory and in extension practice) as passive recipients of information. Through public telephones, rural people can initiate contact with extension workers (Hudson, 1984). Low cost, small-scale printing equipment is used by community groups to produce their own newsletters, bulletins or campaigning material (Zeitlyn, 1983). Local radio and television stations not only enable locally relevant material to be broadcast but also give large numbers of local people the opportunity to be seen and heard 'on the air', thus helping break down traditional distinctions between communication professionals and audience. Media, and increasingly telecommunications, can serve as a bridge between groups and communities within rural areas instead of acting simply as a radial link from a central institution to a mass rural audience, thus facilitating a shift from authoritarian to more participative models of extension.

On the other hand, the technology also makes more feasible the *centralization* of media production, data storage and the distribution of information. This can greatly increase the speed and efficiency of information flows, but also offers greater scope for *control* over access to and use of the means of communication. For the ability to store and communicate information confers power: power to determine what information is to be made available to whom, power to influence the perceptions and attitudes of large numbers of people.

Political control is exercised in various ways, from government ownership of all means of mass communication, through vetting and censorship of media content, to the issuing of guidelines within which media producers can carry on their work. A degree of political control is beneficial — in allocating broadcasting frequencies, for example, and ensuring that mass media take due account of the interests of minority groups. But it can also be restrictive in the context of rural extension, for example, when governments are unwilling to decentralize media production and distribution by promoting local broadcasting; or when official language policy prohibits the use of local vernaculars in print and broadcast media.

Economic control is most readily apparent in situations where the means of communication are owned and operated by commercial interests rather than by governments (although even with government ownership of press, radio and television services, space and air time may be sold to commercial interests in order to defray the cost of the service). The recent spread of new communications technology in many developing countries has undoubtedly been led by the demand of urban populations for entertainment: video recorders showing movies are in abundance in many towns and cities, and the output of television stations consists mainly of entertainment which appeals to the urban audience with money to spend on the goods and services advertised by the companies that finance the television stations, either by buying air time for commercials or by directly sponsoring programmes. In the rural sector, a good deal of the agricultural broadcasting is sponsored by agro-chemical companies, which has implications for the kinds of advice being offered to extension workers and farmers. Thus, although the new communications technologies are quickly spreading in developing countries, it is not always easy to bring them to bear on the concerns of rural people and extension agencies.

For both political and economic reasons, non-government organizations (NGOs) do not generally have the same ready access to mass media channels as government extension agencies. Some NGOs, which see the need to communicate with a large audience as an important aspect of their work, set up their their own communication facilities: for example, the powerful radio stations run by the Catholic Church in the Philippines and elsewhere, and more local stations run by mine labour unions in Bolivia. In countries where the broadcast mass media are accessible only to government agencies, NGOs make use of small-scale, local media such as audio cassettes and video in their work with rural people. On a global scale, agencies working in rural areas of developing countries have relatively little control over how communications facilities are used. Project SHARE (Satellites for Health and Rural Education), launched in 1985 by Intelsat and the International Institute of Communications, aims to make satellite communications available to organizations working in rural areas that would otherwise have little access to such technology (McGubbin, 1985).

Current trends in communications technology thus present a wider range of options for media use within rural extension than has previously been possible. Decisions on investment in new communications facilities are often outside the control of rural extension organizations. But within the constraints and opportunities set by available facilities, choices about how communications technology is to be used in extension are seen to be predicated on more fundamental decisions about extension ideology and approach (Haverkort and Röling, 1984).

ISSUES

Whatever the nature of those decisions, a number of issues must be confronted if the potential contribution of new communications technology to rural extension is to be realized.

Access to the Means of Communication
Research and experience have highlighted the uneven distribution of literacy skills, of radios and televisions, of newspapers and magazines, within and between rural areas. Similar disparities are inevitably emerging with the new developments in telecommunications and computer-based information systems, such as on-farm computing and Viewdata in the UK and telephones in rural areas of developing countries. The use of communications technology to support extension work may therefore increase the relative disadvantage of poorer and more isolated groups (Shingi and Mody, 1976). Conscious efforts are

needed both to use those technologies that *are* more widely accessible and to *increase* people's access to those technologies that extension agencies choose to use. Literacy programmes, mobile video units, audio-cassette recorders (e.g. Byram and Garforth, 1979), radio forums, community television sets — these can all increase the access of potential users to media communication. They can also reduce considerably the unit cost, per recipient, of providing information.

Hardware Primacy

There is a tendency for new developments in information and communications technology to become seen as beneficial, even necessary, components for an efficient extension service. As with radio broadcasting and cinema vans in the 1960s, so now with video technology and microcomputers: they are regarded intrinsically as a 'good thing', even before any serious thought has been given to how they are to be used or what the implications might be for staffing, training and organizational structures. The acquisition of equipment often tends to outstrip the agency's capacity to use it effectively. Capital funds can buy the hardware today: it takes time, commitment and money from over-stretched recurrent budgets to train staff in the use and maintenance of the equipment. Mobile video units without petrol or video tapes or trained production staff, printing presses without graphic artists or adequate supplies of paper, and microcomputers with no suitable software contribute little to rural extension.

Before investing in new communications technology, it is important to define clearly how it will be used to support extension, including the adjustments that will be needed in budget and staff allocations. At the same time, it is clear the *existing* communication facilities are often underused. The potential of radio is poorly exploited in many situations. Print media could be used more imaginatively. Visual aids equipment at training centres and local extension offices is often unused because replacement bulbs are not available, or staff have not been trained how to use it effectively.

Improving Media Content

Technology offers the means by which information and ideas can be conveyed from place to place. The impact of that technology depends largely on the nature of the *content* it carries. The quality of the content can only be assessed from the perspective of its recipients. Only if they see the content as relevant, only if they interpret it as intended by those who designed it, can it have the impact it was meant to have. What little evaluation of media content is done shows that these conditions are often not met, largely because the designers of content make incorrect assumptions about those who will receive it (Garforth, 1982a, 1983). Relevance can only be assured by defining the information needed by the potential recipients and identifying suitable concepts, words, images and associations for conveying that information. This can be done by exploring the recipients' current knowledge, attitudes and practice in respect of the topic in question, and their own perceptions of their information needs. The interpretation of media can be checked by *pretesting* material, with a view to revising it where necessary before it is distributed or broadcast. These procedures can be built into the process of media production through formative, or 'action' research.

Pressure on Field Staff

Communications technology can undoubtedly support and enhance the work of field level extension staff: but it can also put increased demands on them. The success of media-based campaigns in Tanzania, Botswana and China has depended on the efforts of large numbers of extension workers in organizing, training and supporting volunteer group leaders. More generally, mass media may raise the expectations of rural people which they will look to extension staff to fulfil; or may prompt technical questions which extension workers will be expected to answer. The potential of local media may only be realized if extension staff develop production skills and are able to allocate time to such activities. The effective use of media in the field requires that extension staff be able to reinforce, explain and demonstrate ideas communicated by those media.

This complementarity of media and extension workers demands that investment in communications technology is matched by in-service training, the provision of reference material for extension workers, in both of which tasks communications technology itself can play a role (Young *et al.*, 1980, Chap. 5), and adjustment to pre-service training curricula.

Return on Investment?

Finally we come to an issue implied by the overall theme of this volume. What return can we expect from investing in communications technology? Research has begun to provide indicative costs of mass media use in extension (e.g., Perraton *et al.*, 1983b, for Malawi). But there are serious conceptual and methodological difficulties in trying to assess the benefits in terms of economic returns. It is less problematic to evaluate impact in terms of changes in knowledge, attitudes and practices, and here the research evidence is encouraging. But even so, it is difficult to separate the effect of media from other influences on rural people's behaviour (such as the availability of resources and the activities of field extension staff); and in any case, as we have seen, the effectiveness of media and telecommunications is heavily dependent on the quality of the content which they carry and on the structures and processes within which they are used.

It is more meaningful to ask how communications technology can be used to increase the *cost-effectiveness* of rural extension. What functions of extension can it help us to carry out more efficiently? How can it enhance the performance of extension workers in their contact with rural people? To what extent can it substitute for relatively scarce and expensive field staff in routine information provision and so release them for the more challenging and creative task of working alongside rural people as they seek to tackle local problems and exploit local potential?

We cannot, then, answer questions about returns to investment, or impact on the lives and livelihoods of rural people, by looking at the technology in isolation from the context in which it is used. Whether or not communications technology, including the more familiar range of mass media and extension aids as well as recent developments, will make a significant contribution to rural extension depends ultimately on our ability to adjust our organizational structures and extension approaches to take maximum advantage of the potential it offers.

15

Extension Methods Involving Community
Organization and Local Involvement

G. CAMERON CLARK

FAO Consultant, Kingston, Ontario, Canada

INTRODUCTION

This paper is concerned with those extension methods most suited to
working with and promoting community organizations and for
securing the active involvement of the local rural population. It attempts
to focus on a consideration of the key issue of how best to secure the
active participation of people in rural development. Which extension
methods are more conducive to promoting people's participation and
which ones directly or indirectly discourage their involvement?

The vital importance of securing the active participation of the
people for the long-term success of any rural development effort is now
generally recognized and accepted. The World Conference on Agrarian
Reform and Rural Development (WCARRD) held in FAO, Rome, in
July, 1979, involving 145 governments, stated in its Declaration of
Principles and Programme of Action:

> Participation by the people in the institutions and systems which
> govern their lives is a basic human right and also essential for
> realignment of political power in favour of disadvantaged groups
> and for social and economic development. Rural development
> strategies can realize their full potential only through the motivation,
> active involvement and organization at the grass-roots level of rural
> people, with special emphasis on the least advantaged, in conceptual-
> izing and designing policies and programmes and in creating
> administrative, social and economic institutions, including
> cooperative and other voluntary forms of organization for imple-
> menting and evaluating them.

Organizations at the local or community level are but vehicles for generating and channelizing the participation of the people along constructive lines determined, generally, by the members themselves.

PAST EXPERIENCE WITH COMMUNITY ORGANIZATION

Extension has traditionally worked through and with local groups and organizations. The need to involve people and to secure their active participation is therefore nothing new to extension. We have excelled at organizing meetings in general, including result and method demonstrations, and special interest groups, both for production and marketing.

But the main question now being raised by development specialists is: Who have in fact participated in and benefited most from our programmes? Have they not generally been the more progressive farmers, those with perhaps above average resources and resourcefulness? Has there not been a close tie-up between those in leadership positions in the local communities and those involved in our extension programmes? And to what extent have the groups with which we worked been perceived by the villagers as being their own groups or our groups? Has there not been a strong tendency on our part to consider these groups as our extension groups? Have we not directed much of our time and effort towards convincing the people to participate in our programmes?

Of course, we usually involved a few of the local community leaders or prominent local producers in committees to help plan or at least to be involved in implementing the programmes, but the programmes were still basically 'our' programmes. And how many of our programmes catered to the needs of the women and youth? Should we therefore have been surprised that very few of the below average or even average rural families and women participated in and benefited from our efforts?

PRESENT RESPONSE: STRENGTHEN THE DELIVERY SYSTEM OF GOVERNMENT

The need to reach the great majority of rural people previously bypassed by most extension programmes is now increasingly recognized both by national governments and by international and bilateral aid agencies. Since the rural poor were discovered by these agencies over the past decade, there have been many efforts made to reach out and down to the rural poor families. Greatly increased funds have been made available for recruiting and training more extension staff, extension systems have been revised, buildings have been built, and physical facilities improved from national level to the field. A chain of command from top to bottom has been institutionalized in many countries bringing discipline and organization to what had frequently in the past been a loosely organized system. Field staff can now be held more accountable to their supervisors through periodic reports, visits, and training sessions. But to what extent the extension staff are accountable to the people they are intended to serve, still remains a question.

In short, much has been done to improve the extension system as a part of the delivery system of government. Great efforts have been made to help governments to reach and teach the rural poor and to secure their participation in our programmes. But, what has been done to help the poor themselves to reach up to take advantage of the many services being offered by extension and other government agencies? Our efforts have been directed mainly towards improving and refining our capabilities to convince the rural poor to adopt improved practices as recommended by us. We have organized them into a variety of groups and associations designed basically to make them listen to us. What have we done to help us listen to them, to help them to meet together, to analyse their own problems, to share experiences, to think for themselves, to reinforce their cooperative spirit, and to become more self-reliant rather than more dependent upon extension? While more numbers of farmers are being reached, is not the quality of their participation declining to one of passive acceptance and followers of instructions? Is this not a serious and damaging trend which will stifle creative thinking and in the long-run make the people more dependent upon government?

FUTURE RESPONSE: STRENGTHEN THE RECEIVING SYSTEM OF THE PEOPLE

It is suggested that perhaps the time has come for extension to start giving much greater attention to helping the rural people, particularly the rural poor who form the vast majority in developing countries, to develop their own self-directing groups and associations so that a genuine and increasing demand for extension services may be generated.

Extension workers have long known that people learn best and apply most quickly that which they have a strong desire and immediate need to learn. Instead of spending so much effort on developing extension methods aimed at convincing farmers to adopt new practices which we consider desirable, is it not more efficient and effective to concentrate this effort on helping groups of farmers to identify their real needs and then to give them the knowledge and assistance they themselves want? By responding to the felt needs of well-organized and cohesive groups, do you not also have a much higher rate of adoption resulting from group pressure? Is it not better, economically and socially, for extension and the nation both, to help ten members of a small group to increase their productivity and income by 10 per cent than one progressive farmer by 25 per cent?

Extension needs to build up the capabilities of small groups and organizations within each community to undertake the process of problem identification, and problem-solving through discussion and analysis of their local situation. This requires extension workers who understand institution building and group dynamics, and who are able to command the confidence of the people involved because of their technical knowledge and social skills. By working through community organizations, extension encourages the average person to become more actively involved in the decision-making process of these organizations, thereby helping to keep leadership more accountable to the general membership.

If extension does not take the lead within governments to foster the development of independent farmer groups and associations which can assume primary responsibility for the development and welfare of their members, will it not collapse under the weight of its own ever-expanding bureaucracy as it strives to reach, teach, and support the ever-expanding rural masses?

EXTENSION METHODS FOR ORGANIZING GROUPS

In order for a group or any organization within a community to function properly, its members must share common objectives and have fairly similar social/economic backgrounds. As there is considerable social stratification in most established communities in developing countries, it is now increasingly recognized that separate groupings are required for each major social level if the maximum participation of the people is to be achieved.

Extension has long recognized the need to encourage the organization of special interest groups on the basis of commodity or function (such as clubs or associations of poultry producers, cotton growers, irrigation beneficiaries, etc.). However, as it has become more evident that these groupings were increasingly falling under the domination and control of the bigger and more prosperous producers, some extension administrators have started to see the need for separate or special organizations for and by the lower-income and smaller-scale producers, whose risk-taking capacity and managerial skills are quite different from those of the bigger farmers.

We now realize that for group action to be effective and sustainable, membership in any group or organization has to be relatively homogeneous, both socially as well as economically. To foster the development of homogeneous trust groups requires special skills in such activities as conducting community and household surveys using a participatory approach in the analysis of the data collected, promoting shared leadership within a group, decision-making by concensus, problem-solving, group planning, group self-study, and procedures for conducting group discussions and group meetings which lead to group action, satisfaction, and increased group self-reliance. Modern extension workers need such skills if the active involvement and participation of the vast masses of rural poor families is to be secured and sustained on a self-generating basis.

THE POSSIBLE ROLE OF INFORMAL GROUPS

FAO's Small Farmers Development Programme (SFDP), now operating in seven countries in Asia, is one example of the contribution which small, informal groups can make to strengthening participatory extension work at community level. It is a special programme exclusively for the rural poor, aimed at fostering their participation through small groups, in economic and social activities of direct benefit and interest to them.

An extension worker, called a Group Organizer, living and working at village level, is the key factor in the process. He or she identifies the poor families, involves them in a process of analysis of their situation, encourages them to organize themselves into small, informal, homogeneous trust groups, initially around some supplementary income-generating activity of their choice. The programme ensures them access to institutional credit and basic extension services against an agreement to help and support each other under joint liability for loans received.

Basically, it is a strategy for building up a 'receiving system' for the rural poor so that they can receive their fair share of the wide variety of inputs and services being made available through the 'delivery system' of most governments. There are presently over 5000 SFDP groups functioning in Asia involving over 50 000 families, and the number is expanding. FAO is now promoting similar programmes in other regions of the world.

Experience shows that the groups tend to start with economic activities which they already know. It is only after they have successfully completed one or two cycles using their known technology that they feel confident enough to risk adopting a higher level of technology. It is at this point in the life of a group that there is an increasing demand for new extension information.

Extension workers serving in areas where SFDP groups are functioning find that the demand for their services increases tremendously during the second and third years of the programme. This demand for extension services is not restricted to economic activities but rapidly expands into areas of family planning, health, sanitation, literacy, improved housing, etc., based on the needs of the group members. Families who previously were so weak individually gain a new feeling of self-confidence and self-worth through the support of their group. By starting with income-generating activities, backed by group action, the programme quickly develops into one of integrated rural development, but started from below instead of top down.

A key extension tool in the development of these groups is the use of the participatory field workshop. Basically, it is an innovational method of bringing together a cross-section of 60 to 80 staff from all concerned rural development agencies from field to policy level plus 20 to 25 representative men and women from rural poor families in one specific area. By working together in small, multi-disciplinary, multi-level groups over a period of four or five days they analyse the successes and failures of their programmes in a limited geographic area, and develop mutually agreed plans of action. The process includes direct dialogues with rural poor families in their homes, and meetings at village level with groups of poor families and with village leaders.

The entire exercise is highly participatory and produces a new spirit of cooperation between government staff and the rural poor based on improved mutual understanding and respect. It has been particularly effective in developing more positive attitudes amongst senior officials at policy and planning levels towards group-based programmes for the rural poor, such as SFDP.

Participatory field workshops have also proved to be a very effective tool for annual evaluation of ongoing programmes. By exposing the findings of independent research workers, done in advance, with the participatory findings of the field workshop, more realistic and accurate recommendations emerge. The rate of adoption of recommendations from the participatory evaluation field workshops has been much higher than from classical evaluation studies. This is largely due to the fact that the field staff who are responsible for the actual implementation of most recommendations have been a part of the discovery of the need and feel committed to correcting the situation. As some representative group members have also been involved in the process, no further extension work is needed to convince them of the need to change.

Experience shows that the members of successful SFDP groups can also be very effective extension workers with other villagers and other groups. The organization of inter-group visits and exchanges between groups in different parts of a country is proving to be one of the most effective group extension methods. While the 'farmer-train-farmer' approach to extension is gaining in recognition, it is also clear that the 'small group-train-small group' approach is even more effective amongst organized, homogeneous groups. But it requires extension staff who understand the dynamics of groups and who are willing to 'lead from behind'. This requires extension administrators and policy makers who are prepared to build up peoples' institutions instead of their own empires.

In Bangladesh, the SFDP groups have slowly formed their own associations and are assuming increasing responsibility for extension services to their primary groups. In Nepal, the SFDP groups select different members to receive training in different subject matter and, upon return, he/she serves as a voluntary extension worker with his/her own group members and even to other nearby groups. This is in sharp contrast with the practice of the Department of Agricultural Extension in Nepal which is to pay selected progressive farmers to serve as field level extension workers; and, unfortunately, this is tending to undermine the spirit of volunteerism being generated in the SFDP groups. This is an example of how some extension methods mitigate against participation of the people in rural development.

ADDITIONAL QUESTIONS

This leads to several further sets of questions which are relevant in considering these issues. First, can the promotion and use by extension workers of independent, small homogeneous groups be classified as an extension method? If so, how may it best be taught to extension staff?

Secondly, can extension afford to take the time to build and strengthen self-reliant farmers groupings and associations as vehicles for extension work, or should it concentrate on organizing its own extension groupings of farmers? How best may extension field staff be encouraged and rewarded for supporting and working through small, self-reliant, independent groups when a more traditional approach would probably produce quicker, short-term production results?

Thirdly, how is it possible to convince governments that it is in their short- and long-term interest to encourage and assist, particularly the rural poor, to organize themselves into small, income-oriented, multipurpose, self-reliant trust groups and associations as a strategy for securing their participation in development? And, how is it possible to convince those governments which are already encouraging the formation of farmer groups and associations of the need (a) for separate groupings for the small farmers and the bigger farmers if the groups are to be used for extension purposes; and (b) to encourage the groups to gradually undertake inter-group activities and to form their own associations at their own pace and according to the felt needs of the primary group members?

Finally, how can international and bilateral aid agencies be convinced of the need to make long-term commitments (8–10 years) for any extension programme involving the formation of voluntary groupings of the rural poor?

16

Labour Intensive Extension

ROBERT BRUCE

Department of Education, Cornell University, Ithaca, New York, USA

Extension systems become labour intensive — obviously — by having a high ratio of extension workers to clients. That intensity, however, may have been achieved in a number of ways, for a number of reasons, and in the attempt to deal with a number of problems.

Further, it may be fruitless to discuss labour intensity if the discussants literally do not know, or do not agree upon, the meaning of the terms they use. There is a serious need for comparable and well-defined data. This paper, therefore, will begin with an attempt at clarification. Each of the terms in the title of this paper is open to several interpretations, and each interpretation, in turn, provides the basis for a very different set of conclusions.

Labour
Whose labour is under discussion? When we calculate the number of extension workers, are we talking of those persons designated as *professionals* in some countries, or do we include *paraprofessionals* as well? If we include both, do we also include *local leaders*, and if we count them, do we count them as part of the extension system or as being among the clientele?

It is not always clear from the available data what definition is intended. One may think, as does von Blanckenburg (1984), in terms of 'contacts between well-trained officers and clients', but the definition of a 'well-trained officer' may vary from place to place.

Data on labour intensity are of little value if their antecedents are unknown. Accepting that, for internal use, both extension workers and audience levels will be calculated in terms meaningful to the project or programme at hand, I would suggest that we should disaggregate *extension worker* figures as far as possible by level of training (degree-holders, diploma holders, etc.) until standard categories can be developed and used reliably.

Labour Intensive

What does it mean to be labour intensive? McCabe and Swanson (1975) list ratios of agents to economically active individuals in agriculture ranging from 1:71 to 1:153 600! Von Blanckenburg (1984), in his survey of country reports from Asia and Africa, found ratios ranging from 1:500 to 1:10 000. Coombs and Ahmed (1974) cite illustrative 1971 ratios ranging from 1:21 in the Baghlan area of Afghanistan to 1:1725 for Ethiopia. The overall average for the United States in 1984 was one cooperative extension agent to just over 640 persons* in the farm population. At what point along that continuum does extension become labour intensive?

In computing the ratio, one must deal with the question of just which values are to be used. The figures for the United States make a good illustration. The clientele figure in the 1:640 ratio cited above consists of the total farm population; the number of farmers is somewhat smaller and the number of farms smaller still. The divisor in that ratio is all county-level extension workers. If district, state and federal staffs were added, the ratio would be changed slightly. If only agricultural agents were included, it would also be affected. If paraprofessionals (equivalent to diploma holding agricultural assistants in many countries) were to be included, the ratio would be altered again.

Some account must also be taken of extension strategies. One guesses that the ratios cited are mostly derived from gross population data. In point of fact, it is common for an extension programme to be selective as to its audience, so that the actual number of farmers with whom the extension worker is realistically expected to deal is considerably smaller.

The organization of the clientele must also be considered. If an extension worker is working directly with ten agricultural merchants, and they in turn affect the performance of a thousand farmers, is the ratio 1:10 or 1:1000?

*This calculation is based on 10 710 county extension agents, as reported by the United States Department of Agriculture in October, 1984, and a farm population of 6 780 000 as reported in the 1983 edition of *Agricultural Statistics* issued by the same organization.

We should report *client* figures in terms of individuals, client units (households, farms, etc.), or client groups (cooperatives, villages, etc.). Where either client units or client groups are used, they should be defined and the number of individuals also reported if known.

Extension

Several very different 'games' are played under the name of extension and each has its own rules and imposes a different understanding of the term 'labour intensive'. Among the more important are:

(1) *Information transmission.* The object of the extension game in some circumstances may be confined to the speedy and faithful transmission of vital information from a data source to a clientele that is aware of its needs and fully capable of translating the data into appropriate action. In many such cases, the clientele will have access to and use high levels of communication technology. Efficiency — and even effectiveness — may call for high numbers of clients per agent.

(2) *Targeted behavioural change.* The object of this 'game' is to get particular practices adopted and others discontinued. Often, little attention is given to changing the farmer's basic capacity to analyse problems and create his own responses, and the result may be a programme that transmits practices as management instructions. Extension labour intensity is primarily a function of the level of supervision needed to deal with the inability or unwillingness of the client to follow instructions.

A truly educational approach may call for more interaction between agent and client, as educational needs are diagnosed and responded to, and labour intensity will be affected accordingly.

(3) *Individual and group development.* In this approach, no particular behavioural outcome may have been identified as an objective of the extension system, and the extension worker serves as resource and facilitator to the achievement of clientele purposes. Relations of trust must be built and complex educational transactions are sometimes involved. In some cases, localized research must be conducted before educational work can be undertaken.

The object of the educational effort goes beyond adoption or compliance, to include understanding and the ability to construct effective responses. Much interaction of a high order may be needed and labour intensity will be a function of clientele resources, since the role of extension in these cases is additive.

These are not the only 'games' played under the rubric of extension, and the purpose of mentioning them in this paper is neither to advocate one over the others nor to describe them completely. It is, rather, to demonstrate that the extension strategy used will have an effect not only on the necessary level of labour intensity but also on the desirable level. As Coombs and Ahmed (1974) point out, 'It would be foolish to conclude ... that the extension service operating in the Baghlan area of Afghanistan — showing in 1971 a ratio of one agent per 21 participant farmers — must be many times stronger and more effective than Puebla's extension service which had only one field agent for every 1048 farmers ...'.

THE COSTS AND BENEFITS OF LABOUR-INTENSIVE SYSTEMS

Labour intensity puts more extension workers on the spot, and, in doing so, allows for three benefits:

(1) closer supervision of clients, with thus a capability for spotting problems, and doing so quickly;
(2) quicker intervention to deal with the problems spotted; and
(3) closer interaction with clients, presumably leading to better rapport and to feelings of support.

These benefits do not come without a cost, however. As Orivel (1983) points out, agricultural extension is generally labour intensive, and less costly means are available. To this end, it is important to be aware of the costs involved.

The primary costs of labour intensity are clear enough; more people in the field means more money for salaries, facilities for them to work from, logistic support, etc. These are compounded, however, by secondary costs — increased supervision and convolution of the system as more bodies are added.

A third, and more subtle level of cost should be considered. In any system the number of competent people available is limited. When increases in labour intensity are achieved by adding extension workers, they can force the system to accept a lower level of competency, with further increases in the amount of supervision needed and, perhaps, lower levels or quality of achievement. This, in turn, can result in loss of respect for the system by clients, with further erosion of effectiveness. The level of differentiation among the clientele must be considered. If the clients are farmers, all at about the same level of development and with similar levels of resources and similar balances of enterprises, greater labour intensity may mean simply adding workers. Increasing intensity in a differentiated agriculture usually means dropping some kinds of farmers or more specialization among extension workers at the local level. If competent specialists are available in adequate numbers and at reasonable cost, there is no problem. If they must be lured away from somewhere else, costs may be high and no overall increase in total output is achieved. If they are created through training, costs will also be high and delivery slow, or quality may be lowered.

These points have another aspect which, regrettably, must also be mentioned. In situations where status attaches more to position and title, degrees held, etc., than to the innate qualities of the individual, the expansion of a system to achieve a higher labour intensity may mean adding extension workers of lower status, thus endangering organizational prestige and making internal communication all the more difficult. If categories of farmers are dropped, they may well be those with the least status and political power.

DECIDING ABOUT LABOUR INTENSITY

Given the above, it is clear that labour intensive extension systems are neither good nor bad *per se*. In some situations, the greater intensity of a training and visitation system can make the vital difference between a programme which achieves the necessary threshold of success and one which does not. In others, the intensity is not needed and, in still others, it imposes too high a cost in supervision or in distortion of the educational purpose of the programme.

This is a problem that calls for analysis and not for the choosing of sides. Decisions about levels and kinds of labour intensity should start with the nature of the programme involved, the kinds of objectives to be achieved, and the constraints under which the programme must be carried out. The decision can best be made on the basis of several interacting considerations:

(1) the type of extension agent performance implicit in the programme objectives, the programme content to be conveyed, and the client actions called for;

(2) the effects of geographical distribution, and the ease of travel and communication;

(3) the level and types of educational and communication technology available;

(4) client capacities and the needs implicit in them, including such factors as level of literacy, access to transportation, ability to work without supervision, etc; and

(5) the particular methodology or programme strategy which has been adopted, and the implications that flow from that.

As the above considerations are raised and traded-off against each other, what might be called an *ideal intensity* may emerge. Clearly, this will have to be modified further by organizational and situational constraints. Funding levels are the most obvious, but availability of workers with the requisite levels and types of skill are another, and still others may be added. The important thing to note is that the ideal level of labour intensity is situationally determined and that attempts at general judgements on the matter are a waste of time.

Discussion

The relationships between extension workers, the purposes of their work, and the methods they use are becoming increasingly complex. In part, this reflects the availability of many new methods, but it also arises from reassessments of many methods which extension workers have used traditionally. The problems, however, remain the same; they refer to the selection, acceptance, and use of appropriate methods, planning the messages to be conveyed, and evaluating the impacts. The discussions concerning extension methods thus considered a number of novel issues as well as many which have frequently been addressed. The need is to use methods which are effective and efficient in relation to changing rural situations. In general, the approach to extension methods should be continuously critical, avoiding prescriptive communication models.

It has been conventional to recognize that extension work which is labour intensive has the advantage of more and closer contacts between extension workers and farmers. However, it usually implies a hierarchical extension organization with the activities which occur directly among farmers being normally left to the most junior, lowest grade workers. Even so, such methods involve a relatively high cost even though often they may be conveying only general information and advice ('blanket recommendations'). Such extension work may be appropriate in certain situations, but there is often a need to consider the greater use of alternative methods — the mass media, and work with groups or informal membership networks. If traditional distinctions are abandoned, then, for example, in certain circumstances television can be almost as personal as face-to-face contact (although it can also be thoroughly impersonal), while direct contact between extension workers and farmers may at times be little more intimate than the relations between individuals and the mass media.

Today, the mass media are characterized by their increasing availability and thus the possibilities of using them in rural areas. For example, in Kenya, where increasing numbers of farmers now actually read, research results (which have often been long neglected) are being communicated to them through the medium of print, and, as they are being found and compiled, they also become the basis from which to produce a range of media treatments as well as manuals. Moreover, many new forms of communication technology are becoming available with the condition that their effectiveness is considerably increased when multiples of media are used in combination (e.g., computers, telephone lines and television receivers being used together to retrieve verbal and visual data, or to enable the development of interactive community networks). Although a growing range of innovations in communications technology is becoming available, it is important, however, not to adopt these possibilities uncritically, or only to seek ways of using a new medium rather than considering carefully how to exploit it and its power to enhance the communication of the content of particular programmes.

Considerable attention was given to the use being made of video in extension work, and several approaches to programming and production were examined. In Zimbabwe, different donor aid programmes have led to a number of separately equipped, dispersed units, though they may remain unused unless sufficient funds are also made available to staff them. Even when used, questions arise as to the relative merits of centralized and dispersed production units; the former may reduce the medium's ability to deal with local situations, while in the latter the standards of production may be lower and expert information may be unavailable. In Nigeria, video is now regularly being used to support T & V extension work. The fortnightly training sessions are recorded on video as a means of reaching field level extension workers and even the contact farmers with, it is claimed, a greater fidelity in the message transmission than occurs in a chain of human communication links. Video is also being used in several areas as a means of ensuring feedback. By taking a camera into the field, farmers can articulate their problems in their situations and terms. Exchanges of information and opinion by video are also being found effective in stimulating understanding and interaction between areas or communities which have interests in common but which otherwise are not being shared. Thus, a growing number of ways are being recognized in which new communications technology can be an integral part of improving the use of groups in extension work.

The discussion of methods which focus on community involvement in extension work drew upon experience derived particularly in four extension programmes: (a) an expanding network of farmers' groups in Benue State, Nigeria, whose initial stimulus arose from research work by rural development workers attached to a Catholic mission; (b) a rural credit programme operated by a commercial bank in India with a priority focus on the poorer farmers; (c) the reorganization of agricultural extension in a region of Mexico, in order to incorporate farmer participation (for example, in identifying problems and potentials, adaptive research, and developing local community plans); and (d) group development within communal grazing areas in Botswana.

Several themes recurred in the discussion. The use of existing groups in extension work may have advantages, but there can also be disadvantages: grafting new functions on to a group may destroy it; or, large sections of the rural population may be automatically excluded from existing groups. The task of mobilizing and organizing groups might be a proper role for extension workers, but some believed it to be a more appropriate task for other rural institutions (e.g., local government) since it is preferable that those responsible for the delivery (or supply) system should not also be required to develop the receiving (or demand) system. For groups to be able to take action, it is important that they have available to them the means to do so in the form of credit or grants. It was also agreed that non-governmental organizations are ahead of government extension agencies in developing the methods of stimulating and supporting group development and also in those of training extension workers in these processes. National governments and donor agencies (such as the World Bank) are now looking to this experience for guidance. While recognizing that large-scale replication of group processes is difficult, especially by governments which wish to do so quickly, the experience which has been gained indicates a number of common features: initial research is necessary within and with local communities; the groups must take all the decisions; failure must be expected in some instances; and the training of extension workers must reflect methods of working with communities (the analyses of situations and problems, listening, and participation).

Work with groups and communities also has implications for

extension management and organization. There is a need for a change of attitude among extension workers and their managers towards this kind of activity. To be able to undertake the work, extension workers should be given appropriate incentives: possibly of greater importance than salaries are the provision of managerial support, adequate technical back-up, mobility, and training which produces a feeling of competence to work with groups and communities. It is also important to select and involve volunteers from the local communities who can act as change agents, who are committed, and who are accessible and accountable to their communities.

There are also a number of implications for the management of extension work in the use of mass media and the innovations in communications technology. The new media offer opportunities for reaching farmers in more cost effective ways. However, in deciding to use them there should be a consciousness of the possible cultural effects when exposing rural people to multimedia and non-traditional message forms. One of the important functions of new communications technology could be to enable the technical support, back-up and training given to field extension staff to be strengthened (using a range of new means from electronic mail to interactive computer-based staff training to data bases being accessible on-line from the field). For example, in the scattered islands of the South Pacific, radio links via satellite now serve administrative, staff training, and information exchange functions for extension personnel. For the managers in the higher echelons of extension organizations, the use of the developments in communications technology imposes new questions and decisions. The effective utilization of these innovations hinges on the abilities of managers to exploit their potential, recognizing that in some respects the new technologies can be replacements for existing methods while in others they are supplemental. In other words, the new developments require them to be capable managers of communication. Generally, since their background is usually in agriculture, they have little in their experience which prepares them for this. Yet, there is clearly a need for them to be able to evaluate and take decisions concerning the new communications possibilities. This refers particularly to the relations between managers and their field staffs, especially how to integrate these technologies into training systems, and to the effects of using the communications innovations on the crucial relationships between extension workers and their clients.

RESOURCES AND MANAGEMENT

17

Accountability in Extension Work

JOHN HOWELL
Overseas Development Institute, London, UK

There are three main ways of examining accountability in the work of agricultural extension agents: as a measure of performance, as lines of responsibility, and as an aspect of the behaviour of extension agents in relation to farmer requirements. This paper is concerned primarily with behaviour but, for the record, a brief outline is given of the two other approaches to the issue of accountability.

The first way of looking at accountability is to regard it as a *measure of extension performance*, taking extension to mean the public provision of agricultural advice and support. Taking the two centrally important issues in public services of *effectiveness* and *cost*, a fully accountable system of extension is one which provides farmers with the services they require and is able to convey new technical information appropriate to individual farm resources. The 'measure' is the degree of effectiveness in meeting these requirements: most developing countries would score poorly on this measure as other papers in this volume indicate. On cost, an accountable system is one which shows relatively clearly the benefits to public investment in extension services: again, this is discussed in other papers, and it is clearly now a major donor concern.

The second way of looking at accountability is to treat it as an issue of lines of responsibility, or *structure*. Put simply, the issue is of the management accountability of a dispersed, mainly unsupervised, field service to its superiors. But the service itself is, in theory, accountable to farmers also. The main issue of structure is how these two lines of responsibility can be reconciled; but in practice, there is rarely a *conflict* of responsibility. There may be circumstances where extension agents have to reconcile their Department's instructions with those of locally influential political leaders: and there may be instances of pecuniary inducements which encourage agents to put their services at the disposal of individual farmers, disregarding formal regulations. But, in the main, the entire structure is weighted so that accountability comes from one direction only, from the employer.

This is particularly evident when we consider the customary way of considering extension services as the base of a pyramid with a Director of Agriculture at the top. Consultants examining extension services almost invariably look to strengthen the shaky foundations of the pyramid by increasing the flow of resources and supervision from above. But what may be needed is for us to turn this pyramid upside down so that, metaphorically speaking, we stand in a farmer's field and ask what advice and services the farmer requires, and build our organizational requirements — and its lines of responsibility — up from the requirements of the user. This is probably the way that a retailer of house-to-house services (such as insurance) would plan a business operation (and would plan a system of staff payments based partly on service take-up).

This reversal is very much easier said than done, but it is useful to bear this fanciful notion in mind when we come to look at accountability as an aspect of extension *behaviour*. We would all agree that extension and Ministries of Agriculture generally are likely to be successful only when they are responsive to farmers' demands. 'Demand-led' research has a much better record than 'supply-led' research as is evident in high potential areas of countries such as Kenya, India and Zimbabwe. 'Demand-led' services are held to be an accurate reflection of the 'market' for technical advice (and thus they are more amenable to cost recovery than advice or services not requested by the user).

This is all very well; but how, in practice, are we to turn agricultural extension on its head, how are we to change the current weighting whereby the employer is the Ministry and it is the Ministry — not the farming community — which exercises powers of sanction and preferment over field staff? How, in fact, can this weighting be transferred so that the user is at the top of the pyramid, and so that the approach of the extension agent is determined by farmer requirements rather than Department directives at virtually every turn?

There appear to be three approaches worth considering. The first approach is to make the farm community itself the employer of the extension agent. Recently, the case has been made for extension staff employed by cooperative societies and also for fee paying extension associations. For both cases there are precedents. The farmer service centre cooperatives in Malaysia come particularly to mind. In these, the salaries of the technical field services are paid for by the users and in theory they are liable to dismissal or promotion through the society. In practice, however, the societies are closely regulated by the government and the payment of field staff is only met through the levy of a general management charge. It must also be noted that these are on irrigation schemes where a degree of regulation over such matters as variety selection, use of water, and even marketing can be exercised. In less structured systems of agricultural production there is much less scope for such arrangements. These arrangements are, in any event, not *direct* employment of the individual agent by the farmers served.

The record of direct farmer-employed extension staff is, as far as I know, very limited. For my part, the only experience I can draw upon is the attempt to recruit village-employed *bwana shambas* in the Tanga region of Tanzania. In the pilot district of Mukeza, each of the six divisions had 10 to 12 *bwana shambas*. These consisted of 32 Ministry of Agriculture (*Kilimo*) staff (covering four villages each) and a further 37 single village-council-employed *bwana shambas* with brief initial training. All *bwana shambas* attended the fortnightly training. In practice, therefore, the system was one of extension auxiliaries.

The record of the system was disappointing. By the end of only the second season, in 1983, only one-third of the village-employed *bwana shambas* were receiving their salaries, and in the villages covered by the auxiliary system, farmers' attendance at meetings was lower than in *Kilimo* extension villages. It must be said, however, that the quality of performance of the *bwana shambas* was seriously constrained by the unattractiveness of the technical recommendations they were putting forward. In response to this, extension work was increasingly concentrated upon the supply of improved planting by *Kilimo* materials which further reduced the importance of village-bound agents.

It would equally be wrong to make too much of a limited experiment in Tanzania at a time of great difficulty for that country, but I must confess that I do not regard the prospect of *village*-employed extension staff with much seriousness: small groups of farmers with a high degree of uniformity in cropping may have a better prospect of success, but the only examples I know of tend to be outgrower-type schemes (e.g., tobacco in Kenya) where 'extension' is tied up with the company provision of inputs and marketing/processing. From this, it becomes possible to introduce forms of payment, or advancement, by results for extension agents.

The second approach is to introduce user charges payable by individual farmers for specific extension activities. In this respect recent developments in UK agriculture are of considerable interest. Within the ADAS (Agricultural Development Advisory Service), the government's agricultural extension service in England and Wales, there are four sections: veterinary services, land and water, agricultural science and agricultural services. In all but the agricultural services there is already a system of charging for particular activities such as soil diagnosis, treatments, etc., and there is now a proposal, instigated by the Director General, to introduce principles of the market place and charging into the agricultural service itself (Bell, 1984). The argument is that advice is of commercial benefit to the farm business and advice should be treated just like any other service (and the term advice would cover not only farm visits but also conferences, bulletins, etc.). Reactions to the proposal, particularly within the ADAS, have been sceptical on this principle of charging. It is accepted that agricultural advice can be improved by the test of the market place, but there are a number of reservations about charging which appear to me to be particularly appropriate when considering its feasibility in developing countries.

Broadly there are four reservations; first, much of UK extension work is concerned with conservation and environment — in other words, 'common goods' rather than 'private goods'. Second, there is the issue of differential ability of farmers to pay for services and the fear that the poorer farmers will be less likely to obtain services than the better-off. Thirdly, there are the arguments that by charging for services there will be less farmer participation in extension work more generally. For example, farmers would be less willing to allow demonstrations or trials on their fields. Fourthly, there is the issue of the financial returns to a system of charging given the likely high administrative costs of collecting charges.

The fact that all these reservations have been raised in an agriculturally rich country suggests that scope for user charges for extension in the Third World are extremely restricted. In my view there is unlikely to be much scope for increasing the accountability of extension staff simply through a single measure such as user charges.

The third approach is one which I favour. This is to recognize the sheer difficulty of turning the Ministry of Agriculture upside down and

to accept that the only way extension agents will become more accountable to their clients is through a gradual, incremental, process of
involving farmers in some of the day-to-day decisions of extension work
which the agents themselves control. Areas of possible improvement
would be fuller discussion with farmers on the purposes of demonstration plots and, through this, a discussion of alternative sitings, sizes
of plots and any adaptations to the demonstrations. Often these sorts of
considerations are secondary to simply getting a fixed number of plots
established to meet Departmental instructions. In Nepal, this form of
'targetting' is a particularly strong disincentive to consultation with
farmers, and as a result a good deal of confusion has arisen in hill
districts over the differing nature of trials, demonstrations and seed
multiplications. Other areas of possible improvement involving
'accountability' are selection of farmers to attend special training. Apart
from improving selection, this could also ensure some interest among
non-attenders in the technologies learnt by the returned trainees. Even
more rudimentary is discussion with farmers on the itineraries of
extension agents (including their meal times and safari accommodation).

There is still an extraordinarily long way to go in discussing such
matters with the users of extension services, and this will take time for
two reasons. First, there is in many countries simply insufficient interest
among farmers in the work of extension for them to want to become
involved in such details of work organization. Second, in many rural
societies any advance that can be made in involving farmers in such
decisions has to be checked when it becomes evident that a small
number of influential members of the local community are pre-empting
all the services the extension agent is able to provide for their own
benefit.

In summary, the concern in this paper has been with conditions for a
closer relationship between farmer and extension worker. The Ministry
structure itself presents a major problem for re-orientation and it is not
realistic to see any major scope for either farmer-employed extension
agents or direct-user charging. If there is to be any experimentation it
will involve small confidence-building steps involving discussions over
which the extension agent has some discretionary authority.

18

Basic and In-Service Training of Extension Staff

WILLIAM CHANG

Department of Agriculture, Kuching, Sarawak, East Malaysia

INTRODUCTION

The training of staff is one of the prerequisites for an effective extension service. Most developing countries have now developed some form of staff training programme for strengthening the technical and operational capability of the extension service. However, many of these training programmes are not without problems and shortcomings. In this paper, some of the important issues and problems concerning basic and in-service training of extension staff at various levels are raised.

The extension organization of many developing countries has three levels of extension staff: the executive level, in-charge at the headquarters or regional level; the middle supervisory level, in-charge of an administrative unit between the local or regional level; and the field level, as grassroots agents directly in contact with the clientele (the farmers). Two types of training are essential for producing capable extension staff at these various levels — the basic, pre-service training, and in-service training.

BASIC TRAINING

The basic training of extension staff can be classified into three main levels:

(a) *degree level*, the apex of education and training for preparing professionals at the executive level;

(b) *diploma level*, a lower level of education and training for preparing those who are to be responsible for the supervisory functions or the middle level managerial personnel; and

(c) *certificate level*, the lowest level of education and training for preparing the grassroots level extension workers.

In developing countries, agricultural education and training at these three levels are provided through a variety of institutions and services which are frequently under different ministries and/or departments. There often exists little or no coordination among these agencies in tackling the problems of shortage of adequately trained manpower for agricultural development.

One of the problems associated with this lack of coordination among agencies is that the curricula at the degree and diploma levels do not adequately prepare the graduates for employment as executive staff and supervisors in the extension service. One criticism is that too much emphasis is being placed on theoretical training while insufficient emphasis is given to practical training (Maalouf and Contado, 1984). As a result, when these graduates join the extension service they are not able to offer down-to-earth practical recommendations to the grassroots extension workers. Another criticism is that the degree and diploma courses usually concentrate on general agriculture and there is inadequate emphasis on extension education and other related social science subjects. It has now been generally recognized that the various agricultural, social and economic problems of the farmers are interrelated. An extension programme would not be effective in resolving the problems of the farmers if it only deals with technical agricultural problems. The executive and middle level extension personnel who are involved in the planning and supervision of the extension programme, therefore, must be adequately equipped with a knowledge and understanding of the principles of extension, farm management, rural sociology, marketing and other related subjects so that they are able to do their jobs effectively.

These shortcomings in the curricula at the degree and diploma levels have also given rise to similar problems in the pre-service training of the grassroots extension workers at the certificate level. A number of those who graduate at the degree and diploma levels become trainers at the agricultural institutes or agricultural schools which are conducting the certificate level of training. The type of education and training these graduates at the degree and diploma levels have received will inevitably affect the character and quality of the training given at the certificate level; it will remain general and theoretical without much relation to the ground situation in which these future grassroots extension workers will be working. Moreover, the training will also be deficient in extension and other related subjects in view of the inadequate training of the trainers in these subjects.

With the increase in the level of education and the length of experience in operating an extension service, a number of developing countries (which include Malaysia) are recruiting trainees who have attained at least the standard of the General Certificate of Education (GCE) 'O' level or its equivalent for the certificate level courses. In fact, two decades ago Savile (1965) advocated that this should be the minimum educational standard requirement for a grassroots extension worker. However, in order to produce technically competent extension workers, it is my view that the trainees possessing such an educational standard should have at least a good pass in general science and mathematics. This is to ensure that they are able to do well in the technical agricultural subjects at an agricultural institute or school of agriculture. The question is whether or not, in developing countries, many secondary school students who have such qualifications are motivated towards agriculture and life in the rural areas. The observations in Malaysia, for instance, are that those who have done well in the science subjects at the GCE 'O' level are more interested in proceeding on to higher studies or looking for other types of jobs which are urban-based. Usually the candidates we have had for the certificate level courses are the ones who are on the lowest rung of this educational level and who have no other choice of profession.

IN-SERVICE TRAINING

The deficiencies in the basic training of all levels of extension staff necessitate the provision of in-service training for improving the capability of the staff. In-service training is also essential in view of the changing farm technology, the system of operation, and the approaches and techniques used in extension.

The first group of staff to be considered for in-service training are those at the executive, supervisor and trainer levels, because the effectiveness of the grassroots extension workers hinges on the quality of the staff at these levels. In improving the technical capability of the staff at these levels, the training is normally conducted within the country as the technologies are usually generated by local research institutions. Facilities and expertise are usually available in the country for such training. However, in the case of new extension approaches, procedures and techniques, and other aspects of extension, facilities and expertise may not be available locally, especially in those young developing countries which have a short period of experience in operating an extension service. In this case, overseas training or foreign technical assistance may be necessary for the in-service training. Overseas training may be in the form of study tours, short courses and post-graduate courses depending on the need and the subject concerned. As it is usually expensive, it may be beyond the means of the extension organization of a developing country to sponsor a large number of staff for training. One way to reduce the cost is to send a small number of staff for overseas training who, on their return, will disseminate the new knowledge to other staff by conducting seminars or courses. Another way is to invite trainers from accredited institutions to conduct training in the country. In this way, a large number of staff will be trained at a much reduced cost.

There are now a number of overseas training institutions which have courses specially tailored to the needs of the developing countries. However, in view of the unique conditions of each developing country there is a need to adapt the knowledge to the local situation. If the trainees have limited field experience, they may find it difficult to adapt the new knowledge to the local situation. The selection of suitable candidates especially with adequate field experience will therefore be crucial to the effectiveness of the overseas training.

Many developing countries provide regular in-service training for the grassroots extension workers. A number of countries (including Malaysia) have adopted the Training and Visit system of extension (Benor and Harrison, 1977) in which training sessions on specific technical agriculture and extension techniques are conducted fortnightly for the field level extension workers. Regular in-service training is necessary for up-dating the extension workers with the latest technologies and certain aspects of extension work.

Two categories of field extension workers are found in many developing countries. One consists of those who have been recruited in the early days of the extension service but who have not received proper basic training, usually having a low level of education. However, they do have a considerable length of experience in the field. The other category comprises the recently recruited extension workers who usually have a much higher level of education, but have relatively little field experience. The training needs of these two groups of extension workers are therefore different.

In up-dating the field level extension staff with the latest technologies through in-service training, a crucial question which arises is whether or not the technologies introduced to the staff are appropriate to the needs and socio-economic circumstances of the target farmers. In order to ensure that the technologies generated by research are appropriate, the in-service training (being part and parcel of the extension system) must permit a two-way flow of information between the research workers and the field. This two-way flow of information can be brought about during the training by a dialogue approach, in which there is no difference between the trainers and trainees in so far as they are both learners. This approach rejects the assumption of conventional education and training that the teacher knows better; it is what Chambers (1980b) has advocated for the development workers dealing with the rural poor — 'learning from and working with'. It recognizes the value of knowledge of the grassroots extension workers who have direct contact with their clientele (the farmers) — their knowledge of the socio-economic conditions of the farmers, the problems, knowledge, attitudes and practices. In this approach, thus, the trainer can benefit from the experiences and knowledge of the field extension workers who are the trainees. By encouraging the trainees to contribute their knowledge and experiences in the training, it can boost their morale and increase their confidence in their work. Moreover, this approach can make the grass-roots extension workers understand that they too could learn from the farmers. This dialogue between the trainer and the trainees can be brought about in well-planned workshops, group discussions and seminars, where the trainer acts as the facilitator.

Practical field demonstration of the new technologies introduced should still be emphasized in this form of training. This is to ensure that the trainees have practical experience of new technology. Any modifications or adaptation to the local conditions could then be discussed and decided on. Such training would be more beneficial to both the research and extension organizations if the relevant staff from these two agencies were participating together in the training sessions. In-service training in this way can provide an effective mechanism for strengthening the extension–research linkages.

This brief discussion has not addressed itself to many issues related to extension staff training. For example, how can linkages be established amongst different agencies responsible for training to produce the kind and quality of staff needed for agricultural development? In many developing countries, as basic training is inadequate, in-service training is essential for improving the capability of the extension staff. However, a number of countries lack the full range of expertise and facilities required for conducting training in extension and other related subjects, for which foreign technical assistance may be necessary for up-grading and up-dating the staff. Would it be useful to set up institutions for extension research and training on a regional basis to cater for the developing countries of similar background?

19

Manpower Planning and the Development of Natural Resources in Developing Countries*

JOHN FYFE[†]

Overseas Development Administration, London, UK

Identifying and planning the development of skills to relate to the requirements for the extraction and development of natural resources provide one key for the relative success or failure of developing countries to sustain acceptable standards of living in the future. However, despite the importance of manpower planning and development as a subject area, and the critical importance in developing countries of their own natural resources for supporting communities, it is amazing that little attention has been given to how these two subject areas can be brought together to achieve developmental objectives.

In this paper, a short summary is given of the main issues involved in any manpower planning process together with an exposition of the 'diagnostic' approach that I have developed over the years. After that, I will suggest how these notions might be related to the field of natural resource development so as to make a start on bringing together these two very important areas of interest.

Manpower planning and development defies any precise definition. Many disciplines or facets of life have a part to play — economics, sociology, religion, culture — and experience; but basically, manpower planning and development concerns the role of people within an organization or society.

Essentially, the notions associated with manpower planning and development are very simple. However, it is not so easy to identify the

*This paper does not necessarily reflect the views of the ODA, and should not be quoted as such.
†John Fyfe is also visiting Professor of Manpower studies at Manchester University, UK.

225

notions involved or, in fact, to agree over their significance. Manpower planning and development is about a strategy or a process to ensure that any enterprise, or society, obtains the right number and kind of people in the right places and at the right time. This is the simple notion but there are inherent conflicts both in terms of different organizations and levels within society, and different views about what is 'right' or 'appropriate'. In a very narrow sense, manpower planning simply means some kind of replacement planning in terms of the organization and/or society for its immediate requirements. The main stages involved are shown in Fig. 1. In this sense, it would encompass traditional analyses of labour turnover, recruitment policies, career planning for individuals, the relationship between pay policies, recruitment and the deployment of labour, the procurement, selection, placement and training of personnel and, of course, the question of an appropriate return to that labour for services rendered. At any level in society, these simple notions can be portrayed within mythical models that can mean little in practice. In such a framework, the subject matter

1.   DEMAND
                  FORECASTS OF LABOUR/
                  SKILL REQUIREMENTS

                  _____

2.   SUPPLY
            FORECASTS OF LABOUR AVAILABILITY

                  _____

3.   MEASUREMENT OF 'GAPS' NUMERICAL
                  SHORTFALL/EXCESSES
            BETWEEN DEMAND AND SUPPLY

                  _____

4.   ESTIMATES OF QUANTITATIVE
            ADJUSTMENTS IN RECRUITMENT,
      WASTAGE, TRAINING, INTENSITY OF WORKING, ETC.
            TO REDUCE/ELIMINATE GAPS

Fig. 1. The main stages in manpower planning: the myth of a manpower equilibrium.

unfortunately has been seen far too often in a very narrow light and consisting primarily of data collection and analysis.

However, this is a restrictive view about manpower planning and development which is outdated and inappropriate to planning needs for the present and foreseeable future. It is useful to adopt the economists' framework for looking at resources. People are an economic resource and essential to the development of any economy. Entrepreneurial skills of one kind or another deploy human resources. Rewards are required to act as some kind of bridge between the requirements of organizations and/or governments to deploy human resources, and the aspirations and/or willingness of people to offer their services. The context within which that reward system operates can usefully be thought of as a labour market.

All manpower and employment issues have something to do with the relationship between the economists' notion of supply and availability of people, and the demand for their services. Economic theory would suggest that a system of wages, or of rewards, provides a mechanism by which adjustments in the balance between supply and demand are influenced, and that a theoretical notion of equilibrium will prevail after the usual 'caveats' and 'assumptions' are specified, and the notion of the 'long-run' is introduced. This is frustrating for the policy maker and manager alike in any real situation.

Economics can *only* provide a starting point for the subject matter — but it goes *no further*. Manpower planning and development is, in fact, concerned with people in ever-changing circumstances who, individually and collectively, have very different aspirations. People cannot be programmed in any precise way either through a wage mechanism or any other economic force. There will *always* be some degree of imbalance between the 'supply' and 'demand' for labour. Manpower planning and development, therefore, if it is to be effective, must be concerned with an understanding of the 'degree' to which such imbalances 'exist', and of equal importance, the 'reasons' for such imbalances.

It is this diagnostic notion that is most important from the policy angle. All too often in the past — and at present — policy makers in different organizations, and nationally, are purporting to pursue manpower planning and employment policy objectives through the use of a whole range of mechanisms. However, these mechanisms or instruments of policy frequently have very little to do with the 'causes' of the imbalances and problems being experienced in the labour market.

To illustrate: if people belonging to a particular group are not able to offer their services for a particular kind of manual work because of their religious beliefs, then adjustments to the wage mechanisms may not be the most effective means of increasing the supply from that group or society. The list given in Fig. 2 sets out a number of factors that may be principal causes behind imbalances experienced in the labour market.

1.   Slowing down of economic activity associated with the decline of overseas markets and the effects of international recession. Limited growth of employment opportunities in both the formal and informal sectors.

2.   Mismatch between output from the education sector and the 'skill' requirements of the labour market. Co-existence of excess labour supplies and scarcity of skills.

3.   Lack of any definition of the skills required for particular projects and/or natural resource sector.

4.   Inappropriate educational and training outputs.

5.   Inappropriate pay structures in society.

6.   Social demands and pressures that are inappropriate and sometimes inflexible and certainly out of step with economic realities.

7.   Pressures from political parties and trade unions for meeting such unrealistic objectives. Frequently, the so-called political 'drain pipe' situation.

8.   Changes in migration flows.

9.   Changes in age and/or employment structures.

10.   Lack of any manpower planning and/or development activities within a particular project and/or sector, and hence inappropriate recruitment, retention and/or redeployment policies and/or actions.

11.   The effects of expatriate manpower and donor activities — uncoordinated and frequently inappropriate to the priorities for resource development.

12.   Influence through private sector investments and distortions that might be induced, and inappropriate longer term developmental needs of the country.

13.   Misallocation of skills that do exist between 'productive' and non-productive activities within the society.

Fig. 2. Possible reasons behind restrictions/rigidities.

Each of these are taken from real studies. The notion here is that policy makers need to know and understand much more about 'why' imbalances exist before considering which particular instruments or adjustments might alleviate the problem.

This 'diagnostic approach' is not only relevant at the level of the organization but also for purposes of national manpower planning and policy. Admittedly, each of these causal factors interact with each other, and it is not always possible to put different reasons into clearly defined boxes or to give a precise weight or value to each. However, it is important, especially at the micro level, to give some identity to the specific factors that are major reasons behind the imbalances. At the macro level it is important to build up a knowledge about the extent to which such factors are significant in different situations if any attempt at national manpower planning or developing a national employment policy is to be meaningful.

Manpower and employment issues are going to be at the forefront of policy-making for the foreseeable future. The imbalances are not going to be corrected by sitting back and waiting for some new international economic recovery of a scale that will take up all these manpower and employment issues. This means that governments and policy makers at all levels will, for social as well as economic reasons, be looking for more selective, effective and interventionist measures of one kind or another.

One thing we do know from experience is that magic wands and 'instant' packages for removing the rigidities and/or problems are not available. Frequently, improvements can be brought about by systematic examination and consideration of the factors involved and through much hard work and cooperation between people within different organizations. A great deal can be done through improvements in manpower management to reduce the effect of some of these problems. However, it must be recognized that improvements in planning and man management cannot relieve all of these problems. The diagnostic approach can help any organization or country to improve or reduce the scale of the imbalances.

However, the approach also helps one to see areas where shorter-term or even medium-term policies cannot have any real effect. Even defining the areas and/or magnitude of imbalance requiring longer-term and/or more radical solutions is useful. Indeed, the evidence suggests that the very concept of 'employment' in the foreseeable future must be the subject of fundamental rethinking in the national and international context. Increasingly, for this need for more radical thinking about employment, it may be the so-called 'developing countries' that have something to offer the more restrictive notions of manpower planning and employment which prevail in the so-called 'developed' or 'industrialized' countries.

This diagnostic approach to manpower planning and development should be at the forefront of all policy work in this area. Whilst data analysis and the presentation of manpower models had their part to play in the 1950s and the 1960s, it was a part which was more to the benefit of the planners than to policy makers, let alone to individuals. In the years ahead, the collection of information from the labour market and its analysis will, of course, have a part to play in influencing policy. However, the real problems facing society and organizations within society are not solely to do with the collection, analysis and presentation of data, but in obtaining a better understanding about people in the context of their environment and the reasons behind the imbalances identified.

HUMAN SKILLS AND THE DEVELOPMENT OF NATURAL RESOURCES

Three issues have to be considered further if some kind of marriage between manpower planning and natural resource development is to take place in a way that can be of benefit to societies and communities.

The first area of some importance is that of being able to identify the kind of skills required to develop natural resources within any particular community or society. This needs to be done in a fairly precise way. It is not sufficient to indicate that for a country's economic development more skills are required in the field of agricultural development nor indeed to suggest that such skills will include managerial expertise, supervisory skills, and of course technical competence. Such words and statements are all too common in reports that tend to be written by those who talk about economic development but who have very little experience or knowledge of what this could or should actually mean in terms of the actual skills or requirements for developing natural resources. The people that are in the best position to know what these actual skills are likely to be are the operational staff with expertise in the subject matter. However, in the past such people have not given proper attention to the identity and quantification of such skill requirements. The problem seems to be that, although many technical experts recognize the importance of manpower planning and human resource development, they have not translated their knowledge into a format that can be used for the appropriate development of educational and training development policies. This is primarily because such experts have not themselves been trained or are necessarily familiar with the techniques that can be used for that process. Neither has one tended to find that expertise amongst those who pontificate about economic growth and development. One particular technique that will become increasingly important in this context is that of 'task analysis'. With this technique the technical expert will need to learn how to analyse and specify the actual functions being performed by people or required to be performed by people in any particular project for the development of the resources concerned. In building up a picture of the tasks being performed the technical expert will then have to specify the 'actual' training required in order to impart or use the capability of local 'human' resources for undertaking such tasks. Of course, there will be areas where such competence may not be transferred quickly or readily, but until and unless more effort is paid to actually quantifying and specifying the sort of tasks and training requirements, then manpower planning in terms of any project or development of natural resources becomes a meaningless concept.

Secondly, the use of the diagnostic approach to labour markets, or rather improving our understanding about why it is that certain imbalances or problems exist in the development of resources, will be critical for the future. It will no longer be appropriate for the agricultural expert or the engineer to leave manpower, training and educational issues to others who masquerade under the banner of economics or educational development. There is a partnership to be struck with expertise from those disciplines and sectors, but at the heart of all development is the extent to which one can motivate or develop people with the appropriate skills. Those involved with natural resource projects will have to make sure that their knowledge is translated into a form that can be acted upon by policy makers in the field of education, economic development and training policies being pursued at all levels.

If there are weaknesses in the management skills required on such projects then more of the training needed has to come 'to' the project and be transmitted through and from within the project itself.

Thirdly, if the problem is one of motivation and trying to get those coming out of higher education and training onto the land then something must be done to pressurize policy makers into rethinking their policies. In some cases this could mean a radical change to existing policies. For example, if it is essential for a country to develop agricultural resources then it must presumably be equally essential for both the educational and training outputs to be geared more directly into the requirements of that sector once they have been identified. This, of course, has implications for resource allocation and implies the need for a much clearer identity of priorities as far as education and training policies and outputs are concerned. And this has implications for all donors as well as training institutions.

If, however, the problem is one of social prejudice then the governments concerned have a longer-term task to perform. Someone needs to take a lead in getting across to people the importance of agricultural and rural developments and the priority that must be given to these sectors in developing countries. It may be that social prejudice has to be removed slowly through the education process, but it may also have implications for radical changes in pay relativities *and* structures within countries. In many developing countries there are pay policies in operation and the government can have some influence upon pay relativities and pay structures within both the public and private sectors.

If the application of skills in the agricultural and/or natural resource sectors is all-important to a country's future then the rewards for people going into that sector must reflect that priority. In other words, those working on projects using real skills and contributing to real output must receive monetary benefits that reflect this need in a way that will pull some of these resources away from less productive sectors of the economy.

In essence, therefore, there are two critical issues that need to be highlighted in the future. The first is that of a much clearer identity of skill and training requirements, and the second is an attempt to find out more about why imbalances exist — shortages and surpluses — in order that governments and others concerned can take corrective actions that are more appropriate to the often stated 'priorities' for natural resource development.

The diagnostic approach, as an essential element for manpower planning, lends itself to the more pragmatic amongst us. It is a particular approach which does have something to give to the real world and could provide for links to be forged between the technical expertise in natural resource development, the pragmatic manpower planner, and those at all levels of society who are able to influence decisions and policies.

Discussion

In the development of a country's natural resources, a crucial need is to match this development with the availability and quality of the human resources. The methods of assessing this relationship do not have to involve high costs or sophisticated techniques. Two themes which bear on this dominated the discussions on the resources and management of extension services: the practical means of assuring accountability on the part of extension workers, especially to their clients, and ways by which their training could be improved.

The most obvious accountability of most extension workers, given the hierarchical structures of their services, is to their superior officers. Extension workers also have to be explicitly accountable to farmers especially when, as in some Western European countries, they pay directly for information and advice, or where the extension work forms part of the activities of farmer associations. In such cases, farmers recognize that the extension work is a resource which enables them to improve their economic performance, so that (some) extension costs can be recovered. Few opportunities for such developments exist in less developed countries, although it might be possible for farmers to become more involved in making the day-to-day decisions over which their extension workers have some discretion. However, the extension workers are commonly at the centre of a conflict of interests: on the one hand, policy makers and senior administrators require the extension services to assist in implementing national agricultural policy, and on the other, the problems, needs and priorities which farmers perceive. In so far as these are incompatible, the latter are often overlooked when extension workers, in order to reduce role conflict, are being held accountable to the government for achieving national production targets. It was generally agreed that extension services and personnel should undertake the feedback necessary to make policy makers more aware of the problems of rural communities.

In order to increase extension workers' accountability to their clients, ways need to be devised by which farmers could become involved to a greater degree in the planning and execution of extension programmes. The emphasis should be on gradual changes in this direction through the establishment of closer working relationships between farmers and extension workers. These could include feasible local targets being set jointly by farmers and extension workers (e.g., based on trials or demonstrations), the inclusion of farmers on the management boards of extension agencies, the active involvement of farmers in the induction training of extension workers and also, possibly, in annual evaluations of the local extension workers and their activities.

Much could also be achieved through farmer groups (though recognizing that non-governmental organizations are often ahead of government agencies in supporting groups). Local groups could be used to identify local problems (by means of a problem census), to decide on their order of priority through discussion with their extension worker, and to accept work on those problems for which the extension service is able to offer solutions. One example of this kind is in Colombia, where the Instituto Colombiano Agropequario involves groups of farmers in determining the usefulness of research findings and extension recommendations. Another is in Zimbabwe, where a modified T & V system relies on farmer groups playing an active role in determining what advice and services they require, with a high degree of responsibility being entrusted to the extension worker to take decisions with the farmers (which are then fed back and translated into action by superiors in the extension service). Local groups can also be encouraged to initiate the development of their local areas and communities (as in Botswana and Zimbabwe). However, two possible dangers in such developments were noted. First, as farmer groups become more powerful they may be perceived as a political threat by the government. Secondly, as the responsibility for extension work is transferred to farmers, local elites may monopolize the available services.

One related issue is that of disciplinary control within extension services. The deployment of field level extension workers is necessarily dispersed and often they have little contact with, or supervision from, their superior officers. It is thus difficult to check whether or not they are performing the duties expected of them. A rigid structure with a high degree of organizational accountability may therefore be necessary. This occurs in the T & V system, a single line of administrative control with clear responsibilities and schedules of regular visits being given to village extension workers, which also enables supervisors to monitor the work. Farmer groups can also bring pressure to bear on their extension workers if their performance is considered to be sub-standard. User pressures can be more effective and less costly as means of controlling the activities of extension workers than maintaining large bureaucracies to do so. It was agreed that when field extension workers are given clearly defined roles and receive technical and administrative support and back-up, discipline is rarely a problem.

These issues of management relate in part to the training which extension workers receive. In the discussion, several important requirements of appropriate training emerged: for closer relationships to be established between pre-service and in-service training; for a better balance between theoretical and practical subject-matter; for less emphasis on technical subjects in favour of greater attention to communication skills and the principles of extension; and for the training to be related more to the needs of farmers, and also to those of the trainees according to the stage they have reached in their careers.

A problem which is common in many countries is the lack of coordination between the various institutions (schools, agricultural colleges, universities, and ministries of agriculture) which provide the training. One result of this is that the pre-service training is often too theoretical and technical in its content. Thus, entrants to extension services are often poorly equipped in the practical knowledge and skills which their work requires. An exception to this, which was discussed, is in Israel where the government extension service and the Hebrew University of Jerusalem collaborate closely and have jointly established a national Agricultural Extension Centre. This provides pre-service and various forms of in-service training courses, with the latter being tailored in terms of technical subjects and extension education to the experience and needs of individuals. It was agreed that, as a prime requirement, courses and their content should be developed which are not based solely on the needs of the organization but also on those of the trainees and their clients.

In many countries, those who enter extension services are very diverse in their educational backgrounds, ranging from primary school leavers to university graduates. Many of those with relatively low levels of pre-service education lack adequate abilities to observe, measure, and assimilate ideas, so that a necessary part of their training must be to develop competencies in these skills. In doing this, however, the orientation could be excessively technical. It is also essential, as many stressed, to provide trainees with an understanding of human development processes, and of farmers' behavioural patterns, their social systems and their farming systems.

One means of achieving this, which is also flexible, was exemplified by the participatory dialogue approach to training used in Sarawak. In this, prior to attending training sessions, trainees are required to identify farmers' problems and needs, and these are incorporated into the course content. At the same time, trainees are tested to identify gaps in their abilities, and courses are modified in order to deal with these. Such skill-gap analysis and task analysis have been found by the World Bank to be very effective in improving training programmes and reducing costs; they also have long-term implications for national educational policy. More and better training in communication methods is also important so that extension workers are able to use the same language and terms as the farmers. However, it was noted that trainers are often poor communicators, and there is a need to improve their training skills. One possibility is to use specially prepared training packages which, when used by trained trainers, can be employed repeatedly, leading to a rapid diffusion of new skills (an example is the Visual Aids Training Package developed by the ADAS for use in England and Wales).

The training needs of middle and higher level extension staff relate largely to supervision and management. Many countries have the ability to provide this at Civil Service Colleges or Management Institutes, but often these may not be used due to the lack of coordination between government departments. If this and other forms of specialist training cannot be provided locally, one possibility is to arrange for foreign technical assistance to provide short courses in the country concerned. To be effective, this requires close collaboration between the external and local training organizations so that jointly they identify the precise training needs and objectives. Another possibility is to send personnel overseas for training. This creates problems of trainee selection, and of the identification of training needs which have to be matched to the courses available. These problems can be minimized if full account is taken of the likely longer-term career development of potential trainees, if use is made of skill-gap analysis to identify their needs, and when a clear specification of these needs is given to the overseas training centres.

RESEARCH, MONITORING AND EVALUATION

20

Extension Research: Needs and Uses

HARTMUT ALBRECHT
Universität Hohenheim, Stuttgart,
Federal Republic of Germany

A GENERAL FRAME OF REFERENCE

When reflecting on the needs, uses and issues of extension research, a clear understanding is needed of the tasks and goals of extension. From this evolves the 'problems' which need to be analysed. Since the term 'extension' is associated with different meanings, I will define my understanding of it in the following terms:

— goal	to help and enable farm people to change their behaviour in order to solve (or mitigate) their problems;
— means	communication;
— relationship	voluntary cooperation, partnership (i.e. no enforcement, domination or manipulation).

This definition of extension rests on a few basic *functional* assumptions. If we really think in terms of the people we intend to reach, and know their problems, and communicate with them successfully, then solutions which are suitable for and accessible to the 'target' population will be developed, will be adopted, and will solve or mitigate existing problems. Further, the interests of the people are supposed to be the moving force for action towards solving their problems. When the adoption of 'solutions' does not take place then, obviously, some of the assumptions are wrong: we have dealt with 'problems' which are not the problems of the people; we have proposed solutions that are not sufficiently useful for the target groups; or, we have communicated without providing sufficient insight and knowledge about possible solutions.

Extension research, then, should be able to trace the reasons for failures, and to provide hints how to do better under the given conditions or, *ex ante*, how to avoid failures.

GENERAL PROBLEMS OF EXTENSION RESEARCH: THE MULTIPLICITY OF RELEVANT FACTORS IN DIFFERENT SITUATIONS

Whether or not extension can help farm families to solve their problems depends on a multiplicity of situation-specific factors and their interrelationships which, obviously, form dynamic systems, varying or developing permanently in the course of time. From this, four general problems arise for extension research:

The 'Whole' versus 'Parts'

How far should extension research be extended? Is it mainly related to the methods of communication from the different extension systems to the different 'user systems' and their corresponding reactions? Or, should extension research also take into consideration the processes of knowledge generation and the 'fit' of research results (see Fig. 1) (Havelock, 1969; Lionberger, 1982)?

The Detection of Relevant 'Parts' of the 'Whole' in a Given Situation

Even if we restrict research to the complex 'extension-and-user-system', there remains the necessity to concentrate research efforts on those relationships that seem to be most relevant for the existing problem and which can be dealt with effectively. How can this be done in a dynamic system without losing sight of the 'whole', the complex of ecological, technical, economic, social, institutional and cultural conditions?

The Importance of Unintended Effects

Under conditions of many relevant interdependencies we have to admit that extension activities often lead to unintended effects. If so, then how can extension really lead to self-help, without devaluing local knowledge and self-esteem, how can it help to overcome rather than sharpen socio-economic differences (Gotsch, 1972)? How is it possible to avoid the danger that effective short-term solutions may contribute to long-term ecological deterioration, etc? How can we become aware early enough of unintended effects?

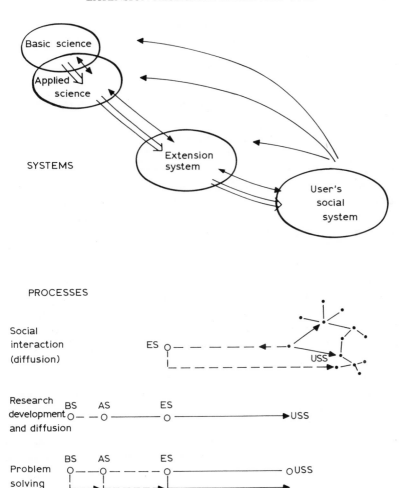

Fig. 1. Generation and diffusion of knowledge (adapted from: Havelock, 1969; Lionberger, 1982).

The Range of Relevance of Research Results, the Problem of 'Similar Situations'

When extension works under varying conditions, some of the research results might be valid in many other situations, while some may only apply in specific situations, 'similar' to that where the investigation has been done. How can one define 'similarity' in practical terms, so that it can be made clear for which situations the particular results can be expected to be relevant? Can we 'accumulate' valid knowledge?

REQUIRED THEORETICAL CONCEPTS AND THE RELEVANCE OF DIFFERENT DISCIPLINES

Answers to the above questions will be reliable only if they are based on useful theoretical concepts. But which ones? There are many disciplines which 'offer' concepts of potential use for extension research (such as learning theory, adult education, communication, diffusion theory, rural sociology, sociology of organizations, role theory, political science, cultural anthropology, farm management, ecology, etc.). But the problems do not sort themselves out according to the existing disciplines. Research based only on concepts from the one or the other discipline is bound to be of fragmentary relevance to the situations of extension work and to the living conditions of the people, and will have little use for solving the people's problems.

The concepts we are looking for should therefore (i) provide insight into the dynamics of behaviour in problem-solving situations; (ii) lead to situation-specific questions (what is relevant here?) which help to construct gradually a sufficiently precise 'model' of the concrete situation (Chin, 1961), the function of which would be (iii) to provide an insight into the 'leverage points for change' (problem-solving through extension and related activities) within the particular situation (Albrecht, 1970).

In this respect, Lewin's field-theory (see Cartwright, 1951) seems to have considerable utility. At the individual level it directs attention to forces that hinder and to forces that facilitate change of behaviour. The 'field-view' of interacting forces can also be transformed into an analytical concept at the macro-level. The 'Intersystem-Model' (Chin, 1961), which is often used, leading to the analysis of interaction within and between the client (user) system and the extension system seems to be especially helpful (Chin, 1961; Havelock, 1969; Lionberger, 1982; see also Fig. 1).

With these concepts in mind one can approach a problem situation in extension from any level (single farm, social system) and from any 'corner' (parts of the user system, the extension system, science system). Being acquainted with a situation to a certain extent we then can direct our additional research efforts to those 'areas', which seem to have high relevance, and we can 'call in' discipline-based knowledge without losing sight of the overall interdependencies.

BASIC INSTITUTIONAL AND METHODOLOGICAL PROBLEMS

Who should do the research in extension? Should it be undertaken at universities, a central unit in one or different ministries, a research department belonging to the most important development agency of a country, or what other kind of institution? In all countries, many vested interests are related to this question.

Thus, in practice, very different answers will evolve, for different reasons. Bearing in mind a number of recurring problems, it seems to me to be advisable to have the agencies of extension research close to the extension activities themselves, hoping thus that real and relevant topics will be investigated. Preferably, such an institution should work as a service institution for the existing extension organizations, not as a control institution. Otherwise, the research process may soon be hampered by resistance, and the research results not taken as a helpful corrective but rejected as invalid judgement and control. It is interesting to note that fruitful ideas in this respect have been available for a long time (for instance, at the Planning, Research and Action Institute in Lucknow), but the practical solutions seldom seem to be satisfying. Is this so and, if it is, why? With respect to social science research work, at least two basic issues seem pertinent.

First, the social sciences have, by their very character, a tendency to be critical, pointing to weaknesses, which seems good to me, if it is done with purpose and responsibility. But only seldom do social scientists commit themselves 'to bring social science research into the mainstream of social action' (Dube, 1981, page 5). The reputation which much social science has earned in the scientific community is that it often results in a neglect of the actual, pressing problems of the people in the field once the immediate research objective (i.e., a research report) has been met.

Secondly, to become more relevant, in my opinion, social science extension research should provide for better understanding of the living conditions and problems of people, be engaged more in action-related research, and ensure that the research results contribute directly to the betterment of the extension activities (Albrecht, 1970; Whyte, 1981). In this respect, there is a great deal to be said in favour of 'participatory action research' (Fals Borda, 1984), although we know that many governmental as well as academic institutions are not fond of it. Therefore, the proponents of action research have to figure out what can be done under given conditions (Röling, 1974).

SELECTED ISSUES

Having attempted to analyse basic problems of the range, methodology, and of the relevance of extension research, a few notes may be added on some selected issues.

Living Conditions of Rural People
The relatively little success in development activities, and the experience that research proposals often have not been compatible with real life conditions have led to the revival of attempts to obtain better knowledge of the 'circumstances facing small-scale farmers' (Opio-Odongo, 1985). New terms reflect this orientation quite vividly: the 'farming systems approach' to overcome the partial view of specialized disciplines, and the interest in an appreciation of 'indigenous technical knowledge'. This reorientation is urgent and provides hope. But the difficulties to be faced in putting it into practice are noteworthy. There is the 'old' problem of how to overcome the difficulties of interdisciplinary cooperation. But 'at the root of the problem lies the fact that officials — agricultural extension staff, planners, research workers, "experts" and others — depend on scientific knowledge to legitimate their superior status. They thus have a vested interest in devaluing indigenous technical knowledge and imposing a sense of dependence on the part of their rural clients' (Howes and Chambers, 1978, page 7).

A very relevant part of knowing the conditions of life of rural people is to know the living conditions of women. Recently, it has become a topic of open discussion that many development projects have had detrimental effects for them (Jiggins, 1983b). While the issues in this field of research are clearly formulated, the main problem at present seems to be how to put the conclusions into practice. Various proposals have been made (Collinson, 1981; Chambers, 1983), and experience should be exchanged intensively (Grosser and Ibra Ba, 1979/80; Cernea and Guggenheim, 1985) and, hopefully, applied.

Communication
It is quite obvious that extension can help to solve problems only if

successful communication takes place with rural people. Time and again failures have been reported in inter-personal communication (Turnbull, 1963; Wallman, 1965; Fuglesang, 1973; Ascroft, 1974), as well as in media-based activities such as Development Support Communication (Ugboajah, 1972; McAnany, 1980). But while the farming systems approach seems to be widely taken up, a similar development with respect to the analysis of communication problems does not seem (so far) to be developing. However, more interest in the living conditions of rural people and more involvement and participation will, one would hope, also increase the interest and practical activities towards an improvement in the field of communication (AERDC, 1976; Kidd, 1982). Can gaps in communication partly be 'bridged' by 'paraprofessionals' who are close to the target population and the extension-workers (Esman et al., 1980)?

Organization and Management of Extension Services

Organization and management to a high degree determine the working conditions of all involved in extension work. Therefore, this is an old issue. It has again become a general concern since the World Bank took over the so-called Training and Visit approach (Benor and Harrison, 1977; Benor et al., 1984) as a general orientation for its projects. This approach originally was developed for irrigation projects. Its widespread application in rainfed agriculture in Asia and recently in Africa has brought about reactions ranging from praise to sharp criticism. The problems are mainly seen in a too rigid visiting schedule, a generalized pattern of organization and management that does not sufficiently consider the wide variety of relevant situations, and the high costs of the system (von Blanckenburg, 1982; Howell, 1982b; Scoullar, 1984). More fundamentally, many (Coombs, 1980; Röling and De Zeeuw, 1983; and others) criticize that it is basically a 'top-down' approach, which gives it little chance of really dealing with the problems of the rural poor; rather, it empowers an agency outside of the influence of the target population, while hindering the needed participation and mobilization.

Although it would be difficult to attribute helping or detrimental effects to the approach and to its elements *per se*, due to the inter-relatedness with many other factors, the debate on the T & V approach has given a new emphasis to the need to analyse the broad and difficult field of organization and management of extension services (Cernea, et al., 1983). This is true also for the issue of how to 'monitor' on-going activities and reactions to them, as well as for the question of evaluation (Cernea and Tepping, 1977; Ay, 1980).

21

The Monitoring and Evaluation of Non-Material Objectives of Extension

PETER OAKLEY

Agricultural Extension and Rural Development Centre, University of Reading, UK

Rural extension in its broadest sense is not limited to the process of technological and knowledge transfer in the agricultural sector. Reference has already been made in this volume to the evolution and diversification of rural extension into such non-agricultural fields of home economics and nutrition. In the past decade or so this evolution has taken on further dimensions and we have begun to see approaches to the tackling of problems in rural areas which are not, in the first instance, limited to physical manifestations of these problems. A new *genre* of analysis of rural problems has led us into such areas as local political structures, institutional arrangements for the distribution and control of resources, and the relative access of different rural groups to these resources which are indispensable if such groups are to benefit from technological development (see, for example, de Kadt and Williams, 1976; Pitt, 1976; Elliot, 1977; Kitching, 1982). We have also begun to question whether rural extension is the sole preserve of formal, bureaucratically structured extension services, and to accept the widespread activities of non-governmental services as equally legitimate extension practice.

Extension practice, therefore, which does not immediately promote the transference of technological information, is concerned with a different range of objectives. A review of extension practice, principally in the non-government sector, reveals the following as the kinds of *objectives* of such practice:

(i) to promote the *participation* of rural people in projects and programmes designed to assist their development;

(ii) to develop the *organizational base* of rural people;
(iii) to create *awareness* among rural people of their problems and to build up *solidarity* to tackle such problems;
(iv) to encourage greater *self-determination* and *self-reliance* in order that rural people might assume responsibility for the direction of their own development.*

Extension programmes and projects often contain references to similar objectives, but usually they are not included in programme evaluation which invariably concentrates upon the tangible or *quantitative* results of the extension activity. Essentially, the above kinds of objectives are *qualitative* in nature and, as such, their effects cannot be completely understood by conventional monitoring and evaluation (M & E) techniques. In dealing with the material objectives of extension activities, evaluation is concerned with *results* which are quantitative and which can be expressed in some statistical form: in dealing with the non-material objectives, we need to understand the *processes* which are taking place and which are essentially qualitative in nature, and which cannot necessarily be measured in quantifiable form. Non-material objectives, therefore, demand a different approach to M & E. More conventionally M & E is concerned with the *measurement* of results on the basis of which some form of *judgement* is made on the likely effect of the project. In terms of the non-material activities of extension projects, M & E will be more concerned with *description* which will be *interpreted* and thereby provide an understanding of the effects which have occurred (Oakley, 1984). Thus:

Material Objectives $\longrightarrow$ Measurement —— Judgement
Non-Material Objectives $\longrightarrow$ Description —— Interpretation

It should be noted that in terms of the M & E of the non-material objectives of extension practice we are still very much at the stage of exploration. Whilst an increasing amount of extension literature sees non-material objectives as legitimate concerns of extension practice, we are only beginning to understand how we might explain the effects of such objectives. The work of Haque and his colleagues (Haque *et al.*, 1977) in broadening the basis of evaluation criteria was the first useful

*The range of objectives is drawn from a study of over 100 projects which had received support from OXFAM. For a fuller discussion on the nature and methodology of these projects, see Oakley and Winder (1981).

framework for the M & E of such objectives. Huizer (1982), in his examination of the Small Farmer Development Programme in Nepal, developed the framework further and suggested the kind of criteria which could be used to understand both the material and non-material effects of that Programme.

In order to bring more substance into our examination, we can take one non-material objective — 'participation' — and consider the question of its monitoring and evaluation. Since the 1979 World Conference on Agrarian Reform and Rural Development, the 'participation' of rural people in extension programmes and projects has become a commonly expressed objective.* This 'participation' is not defined merely in terms of direct participation in the economic benefits of an extension project, but also in terms of such intangible processes as participation in decision-making, project-planning, and in the evaluation of project activities; as examples of the voluminous body of recent literature on the concept of 'participation' in development see: Pearse and Stiefel, 1979; Uphoff et al., 1979; Oakley and Marsden, 1984; Bhaduri and Rahman, 1982. Participation and extension is a phenomenon which occurs over time and it cannot be measured by a single camera 'snapshot' kind of exercise. Participation is a process which unfolds throughout the life of an extension project (and which also continues when the project formally ceases), and it has a range of properties and characteristics. The M & E of participation in extension involve a number of measurable aspects (e.g. numbers attending extension meetings or levels of credit use and repayment). Equally importantly, however, we will need to be able to understand the unfolding, intangible process of participation and to interpret the wide range of results which might occur.

The M & E approach to participation has a number of important elements:

(a) *Base Line Survey.* M & E involves comparative judgements of situations *before* and *after* an extension intervention. It will be

*In 1979 the Food and Agriculture Organization of the United Nations (FAO) sponsored a World Conference on Agrarian Reform and Rural Development (WCARRD). The conclusions of this Conference, which were widely accepted by member nations, have since become the basic guide to the FAO's; support for rural development programmes and projects. These conclusions were presented in 'The Peasants' Charter' which was published by the FAO as the major statement from the Conference (FAO, 1979).

important to structure a base line survey which describes the situation in a particular area in terms of participation before the extension activity begins.

(b) *Group Profile.* It is equally important to have a description of the *qualitative characteristics* of the extension group before activities begin. This will be necessary if we wish to note qualitative changes which occur in the group as a result of the extension activity.

(c) *Group Participation in M & E.* Extension group members should have a role to play in on-going M & E. The active involvement of group members in M & E should thus be built into the M & E system from the beginning.

The initial issue, however, in the M & E of participation is the identification of relevant *indicators* which can be used to describe the process and later interpreted in order to determine the results which have occurred. Such indicators will need to reflect accurately the changes which have taken place; they should also be able to be identified and monitored, while the cost and time involved in the collection of data and information relevant to the indicators should be kept low. There is a substantial body of literature on relevant indicators for the M & E of the material activities of extension; there is much less on what we might term 'social' indicators which inevitably refer to 'measurable' social development activities (e.g., education or health programmes), but few address themselves to the problem of 'unmeasurable' concepts.

We need, therefore, to identify a set of indicators which will reflect the process nature of participation and which will help us determine the nature and extent of participation which has occurred as a result of extension activity. Given the qualitative nature of the process of participation, however, inevitably we are dealing with indicators which similarly are qualitative. Therefore, the need is to identify the kinds of *actions, events* or *changes in behaviour* which will reflect the changes which are occurring. In other words, we need a series of *phenomena* which we can observe, or about which we can record data or activities as appropriate, and which will characterize the qualitative changes which are taking place.

We can best understand this by the following two sets of indicators which could be proposed for the M & E of a process of participation. This material is taken from a recent study (undertaken by the author for the FAO) into the M & E of participation (Oakley, 1984):

QUANTITATIVE INDICATORS

(i) *Group Formation*:

    (a) Number of groups per action area

    (b) Totals of group members per year of formation

    (c) Group members as a percentage of village adult population

(ii) *Participation in Group*:

    (a) Number of group meetings per month

    (b) Number of members attending group meetings (male/female)

    (c) Number of members involved in group activities per month (male/female)

QUALITATIVE INDICATORS

(i) *Group Characteristics*:

    (a) Genuine and spontaneous participation in group meetings and activities

    (b) Emerging feeling of group purpose and solidarity

    (c) Positive and regular interactions between group members

    (d) Awareness of issues and problems

    (e) Enthusiasm and support for group activities

(ii) *Self-Reliance*:

    (a) Groups themselves meet/discuss to identify problems

    (b) Ability of groups to plan and organize activities

    (c) Reduction of group dependence on extension staff

(iii) *Independence*:

    (a) Organizational growth and development of the group

    (b) Ability of group to progress without extension staff

    (c) Establishing contacts with other agencies

    (d) Linking up with other groups

    (e) Formalization of group organizational structure

The above indicators and their characteristics suggest what kinds of data and information could be collected in order to be able to evaluate a qualitative process such as 'participation'. The major problem is to determine how such indicators, and their characteristics, are to be observed and recorded. We are not dealing with the more simple task of measuring, for example, the increase in crop production as a result of a package of inputs. How do we observe and make a judgement on, for example, '. . . an emerging feeling of group purpose and solidarity'? We need to give substance to the qualitative indicators and relate them to some observable activity of the project group in order to be able to understand them in action.

The approach to the M & E of such non-material objectives as 'participation' will be continuous description around a number of predetermined indicators. This description will involve the *observation* of activities and events related to the indicators and the *recording* of data and observations for future interpretation. In terms of the process of continuous description, the following are suggested as appropriate forms for collecting relevant data and information: project group *log books*, in which group activities, events and actions can be recorded and accordingly interpreted (a critical issue in the keeping of such log books is the role the group will have in this task); *extension agent's diary*, in which the agent maintains a continuous record of observations and descriptions in terms of the indicators already determined; and the use of *case studies* or individual histories of a particular project over time which can be used to illustrate the process of participation as it unfolds. Given the qualitative nature of the process of participation, its evaluation cannot be undertaken solely by an external extension agent. In recent years the relevance and practice of participatory evaluation have achieved increasing relevance, particularly in terms of our understanding of non-material objectives. The literature and our knowledge of the practice of participatory evaluation is now quite considerable, and its techniques are entirely relevant to the evaluation of participation.

The final major issue is that of *interpretation*. Given the fact that much of the material we will gather will be qualitative in nature, its interpretation becomes critical. This interpretation is the measurement of the nature and extent of the participation that has resulted from the project. In terms of interpretation, a number of problems are evident: (a) the subjective nature of the recording and observations; (b) the fact that participation is a slow, unfolding process with the result that events may unfold slowly and the recording and observations may lack substance; and (c) the probability of inconsistency and unbalanced recording with the result that the material available may be patchy and incomplete.

Field work is currently under way in a number of countries testing out quantitative and qualitative indicators of participation and the methods by which the indicators can best be recorded and observed (see, for example: Hall, 1977; Fernandes and Tandon, 1981; Rahman, 1982). Although extension evaluation is still largely dominated by a greater concern to measure the material objectives of extension activities, the broadening of the interpretation of rural extension demands that similarly we should attempt to measure non-material activities. The consequences of extension practice are not limited to tangible, concrete results. Rural extension can legitimately aim to tackle problems of a structural or institutional nature, and correspondingly, can seek to determine the effect of its efforts in terms of such problems.

22

Methodological Issues in the Evaluation of Extension Impact*

GERSHON FEDER and ROGER H. SLADE
The World Bank, Washington, DC, USA

INTRODUCTION

The impact of extension is an imprecise term. Does it, for example, mean the full set of changes, direct and indirect, resulting from an extension initiative, or does it have narrower connotations of change — for example, in the organization of extension or, less simply, the immediate and direct effects on farmers of extension activity. The literature is littered with such semantic niceties concerning the definition of monitoring, evaluation and impact (see, for example, United Nations, 1984), but these need not detain us. The purpose of this paper is to examine some of the methodological issues and problems involved in undertaking an evaluation of extension designed to observe and measure changes in the operational efficiency of extension agents, changes in farm husbandry knowledge, and the increase (if any) in agricultural productivity induced by the introduction into an area, with agricultural extension, of a more intensive system. We do not

*The World Bank does not accept responsibility for the views expressed herein which are those of the authors and should not be attributed to the World Bank or to its affiliated organizations. The findings, interpretations and conclusions are the results of research supported by the Bank (RP0672-29); they do not necessarily represent official policy of the Bank. The designations employed and the presentation of material in this paper are solely for the convenience of the reader and do not imply the expression of any opinion whatsoever on the part of the World Bank or its affiliates concerning the legal status of any country, territory, city, area or of its authorities or concerning the delimitation of its boundaries or national affiliation.

therefore specifically address issues that arise in evaluating the contribution of an established system of extension to agricultural output. Moreover, attention is centred on extension in less developed countries, the locus of much recent investment, rather than the more sophisticated systems of Europe and North America.

A FRAMEWORK FOR EVALUATION

To have a reasonable chance of success any extension system must become well-known to farmers. This is mainly a matter, in most developing countries, of ensuring that there are sufficient adequately trained and equipped staff. It is important, however, that farmers be aware of the availability of these workers and their ability to provide useful information and to answer questions. Once a satisfactory level of awareness is established amongst the farming community, it is reasonable to expect that an effective process of knowledge dissemination will begin (providing, of course, that the extension service offers information to farmers that is relevant to their needs). An increase in farmers' knowledge about crops and cropping practices is the intended direct product of extension. Obviously, managers of an extension system hope that additional relevant knowledge will lead to the adoption of improved husbandry by cultivators and will ultimately be translated into increased agricultural productivity.

Knowledge, once disseminated by the extension service and acquired by the farmer, has a tangible, measurable product only if applied. The application of such knowledge by farmers is generally termed adoption and is usually measured by adoption rates, that is, the proportion of farmers applying knowledge of a particular crop practice that they have acquired from the extension service. Although adoption cannot take place without knowledge, it is not an automatic consequence of the acquisition of knowledge as many other influences may affect a farmer's decision to adopt. Amongst these is the profitability of the practice, the availability of key inputs, sufficient credit, and complementary knowledge. Hence, although adoption is commonly taken to be an indicator of the output or effects of extension (its measurement is relatively easy) it does present significant problems of attribution which can be tackled only with considerable difficulty and complex analytical tools. Nevertheless, the measurement of adoption rates, and their corollary, the reasons for non-adoption, is a legitimate evaluation activity capable of yielding valuable insights for extension management and policy makers.

Similar problems beset the use of crop yields (productivity) as an evaluative measure. Unless data from a complete analytical framework are available, yield data provide only a rough measure of the effects of extension. A conceptually complete evaluation requires that information on the effects of the project be collected in accordance with the schema in Fig. 1.

	Before Project	*After Project*
Without Project	The situation before the time the project is introduced in an area identical to that where the project is planned.	The situation after the project has been introduced in an area identical to that where the project was undertaken.
With Project	The situation in the area where the project is planned before the project is undertaken.	The situation in the area of the project during and after it has been implemented.

Fig. 1. A complete evaluation framework.

Figure 1 demonstrates that an estimation of the effects of extension (or any other project) requires a comparative approach along two dimensions. Rarely is this possible for either practical or administrative reasons. Sometimes, however, it is possible to work along only one dimension, most commonly, studies of the situation in the area where the project is undertaken both before and after (as well as during) implementation. Even this somewhat inadequate approach is not possible in areas (projects) where monitoring and evaluation efforts are not made prior to the project and subsequent implementation covers the entire country or state. In these situations evaluation must remain, as it were, within a single cell of the matrix — the situation in the area of the project during and after implementation. Such a restricted form of evaluation, common in many project situations, is unable to yield definite answers about the effects of extension.

STUDIES IN THE EVALUATION OF EXTENSION

The literature on the evaluation of extension is extensive and has recently been summarized by Orivel (1983). Those who have attempted to evaluate extension have done so from a variety of perspectives with mixed results. Some, usually aggregative, studies (e.g., Evenson and Jha, 1973), recognizing the dependence of extension on agricultural research, combine the two and have shown a positive relationship between them and agricultural productivity. Other studies have concentrated on estimating the effects of extension by comparing farmers who have contact with extension agents with those who do not (e.g., Harker, 1973). These evaluations underestimate the effects by not taking into account the way in which information, once acquired by one farmer, can be passed on to others. Other studies have concentrated on assessing extension performance by measuring the extent of farmer-extension agent interactions (e.g., Chambers and Wickremanayake, 1977; Giltrow and Potts, 1978). Other studies have examined the extent to which extension agents' visits to farmers are biased in favor of the rich and influential (e.g., Leonard, 1973). Some evaluators see the internal efficiency of the extension system as the crucial parameter and study farmer to agent ratios and the quality and motivation of extension agents (Hoeper, 1983) or the location and mobility of extension workers (Rahim, 1966). In the context of exploring the relationship between education and farmer productivity some investigators have sought to establish whether extension is a substitute for, or a complement to, education (Lockheed et al., 1980). Such studies, however, even when they yield unambiguous results, shed little quantitative light on the net benefits of extension. To do so requires not only rigorous comparative analysis but a formal means of establishing whether, if positive effects are observed, the effects are commensurate with the costs incurred to produce them. Such studies have been undertaken in developed countries, of which Griliches' (1958) study of hybrid corn in the USA is perhaps the most well-known. Studies in developing countries have usually been less rigorous, evaluating extension through a simple before and after comparison of crop yields (e.g., Benor and Harrison, 1977, and data reported for Ethiopia in Lele, 1975). Such studies, however, do not identify, and hence evaluate, the contribution of extension to increases in output because they do not separate out the contributions of factors such as material inputs, soil quality, supply constraints and other variables likely to influence output. Nor do they take into account the possibility that, with time, there may be autonomous growth in productivity.

In recent years there has been a re-emergence of interest in evaluating extension partly as a consequence of the recent, but widespread, introduction of an intensive and highly supervized form of extension — the Training and Visit (T & V) system. Well described in Benor et al., (1984), the system requires a high degree of professionalism, relatively high ratios of extension staff to farmers, frequent staff training, the intensive and regular supervision of field staff and a strict adherence, by extension workers, to a predetermined schedule of farm visits to preselected contact farmers. The rapid spread of the T & V system has given rise to claims of tangible success (e.g., Benor and Harrison, 1977), and also provoked considerable criticism (e.g., Moore, 1984). But both proponents and critics lack substantive and objective empirical evidence with which to substantiate their claims. More rigorous evaluation is required. To do so, we chose a comparative methodology based on a case study which, because it was undertaken at approximately the mid-point of an extension project, operated mainly within the two right-hand cells of the evaluation matrix in Fig 1. Nevertheless, as explained below, strenuous efforts were made to approximate the requirements imposed by the two left-hand cells of the matrix.

A CASE STUDY IN EXTENSION EVALUATION

Our evaluation covered several aspects of the extension–farmer–productivity chain, and several analytical approaches were applied to deal with different issues. To clarify and quantify the impact of T & V extension on (i) system efficiency; (ii) changes in the rate of growth of knowledge dissemination; and (iii) farm output, a comparative analysis of extension activity and productivity in two, specially selected, geographically adjacent areas — Karnal district and Kairana *tehsil* in North Western India — was undertaken. These two areas are similar in most agro-climatic respects, but had different extension systems during the period of study. In Karnal, T & V extension was operating, having replaced the older system of multi-purpose extension agents, which continued to function in Kairana. A detailed farm-level survey was conducted in 1982/83 in both areas, three or four years (depending on the season considered) after the T & V extension system was introduced in Karnal district.

To establish that the extension system had indeed been reformed, we asked farmers whether they had noticed any change. Improvements in the operational efficiency of the extension system were assessed by examining the relative importance of farmers' sources of information in relation to their exposure to extension in the two areas. Extension agent interaction with farmers was measured by calculating visit frequencies by farm size, further disaggregated in the T & V area, according to whether the farmer was a contact or non-contact farmer (Feder and Slade, 1984).

During the sample surveys, farmers were questioned on their knowledge of specific practices, and about the time when they first learned them. Knowledge is difficult to measure without conducting a thorough examination of a respondent's understanding of all aspects of a given recommendation. For some practices this was possible, but for others detailed testing was beyond the time and resources available. In such cases, however, it was possible to establish the farmers' awareness of the practice. Such awareness is an important indication of knowledge because, by definition, a farmer who is unaware of a practice cannot be familiar with its detail. From the resulting data, the growth in the number of farmers who were aware of different technologies between 1978 (the year before T & V extension was introduced in Karnal) and 1982 was calculated for both areas. Later analysis employed two alternative standard specifications (logistic and negative exponential) of the time path of growth in knowledge diffusion (Feder and Slade, 1986). By comparing, for the with and without T & V situations, the parameters estimated from these functions we established practice-specific differences in the rate of growth in knowledge diffusion in the two areas.

Subsequently, econometric techniques which accounted for differences in the quantities of variable and fixed inputs, the types of soils, human capital, irrigation (both quantity and quality), and the production environment were used to estimate the percentage output differentials between the two areas for two major crops: high-yielding varieties of wheat and rice (Feder et al., 1985).

Such an approach implies that when differences in variable inputs are controlled, the effect (if any) of extension on input use is ignored. If differences in the use of variable inputs are not taken into account, then the output differential includes the effect of extension on farm efficiency as well as the effect on the use of inputs, provided that price differentials are also controlled. A complete accounting for the effects of extension should therefore take both types of effect into account (a simplified representation of these two effects is depicted in Fig. 2). Our analysis attempted this, but owing to inadequate price data the possible effect of extension on input was eventually ignored. The residual productivity differential between Karnal district (T & V area) and Kairana *tehsil* (non T & V area) not accounted for by the extensive set of explanatory variables listed above could have been attributed to differences in the extension systems if there were no other systematic factors differentiating the two areas, and if it could be assumed or established that the two areas were, in 1978/79 (pre-T & V), at the same productivity level, after similarly accounting for differences in the levels of explanatory variables in that year. This last assumption could not be maintained, and we therefore undertook detailed calculations to establish the baseline productivity differential.

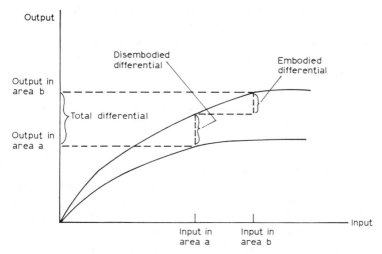

Fig. 2. Embodied and disembodied productivity differentials.

If a similarly detailed farm-level sample for 1979 had been available, we would have replicated for the base year the econometric analysis which we applied to the data from the 1982/83 sample, and the residual productivity differential for 1979 would have been estimated directly. In turn, this would have allowed us to test whether the 1982/83 residual productivity differential was larger than the 1979 differential. As above, any difference between these two levels would then have been attributable to T & V extension work assuming that the rate of increase in productivity between the two areas, in the absence of any change in the extension systems, was the same.

Unfortunately, such detailed farm-level data from our two study areas for 1979 were not available. However, some data derived from seasonal crop-cutting estimates were obtained. These provided a time series of mean yields for wheat and rice for both Karnal and Kairana. These mean yields had a number of deficiencies. First, sample sizes were smaller, and thus the number of observations underlying the Kairana *tehsil* means was only 30–40. Second, the mean yields did not differentiate between irrigated and unirrigated conditions, or between high-yielding varieties and traditional varieties. Our 1982/83 sample, however, focused only on high-yielding varieties under irrigated conditions. Third, no information was available on the mean input levels (and other relevant explanatory variables) pertaining to the sample observations used to calculate the mean yields. Fourth, the mean yields in any given year included random elements which fluctuate over time, e.g., a severe pest problem in a given year, or an adverse micro-climatic condition such as hail. Thus, the differences in mean yields were not directly comparable with our estimates of productivity differentials.

In order to overcome these deficiencies, we undertook a number of adjustments so as to derive from the available data mean yields for 1979 which were compatible with those used in our analysis of the 1982/83 survey data. We then utilized econometrically estimated values of the parameters (based on 1982/83 sample survey data) associated with explanatory variables (e.g., inputs) to generate an estimate of the residual productivity differential between Karnal district and Kairana *tehsil* in 1979. This differential was then subtracted from the one estimated for 1982/83 to obtain the net differential in productivity attributable to T & V extension. The gains due to this net differential were calculated under several alternative scenarios, and subsequently utilized in a cost-benefit analysis of the incremental investment in T & V extension. Only the differential for wheat was used, as that for rice, although similar in size, proved not to be statistically significant.

We originally intended to evaluate the effect of T & V extension work on the adoption of a variety of farm practices and the use of inputs. However, the lack of detailed farm level data on adoption and input use in the pre-T & V (baseline) year implied that the analysis could be carried out only under the assumption that adoption and input use were identical, *ceteris paribus*, in the T & V area and in the control area before the introduction of T & V. This was deemed to be too restrictive, and abandoned. We did, however, obtain estimates of the upper limit of the effect of T & V extension on the use of the main farm inputs and these showed the effect to be small (probably because both areas were agronomically advanced before T & V was introduced in Karnal).

We also intended to specify extension variables at a degree of detail which would have provided operational insights into, for example, the optimal agent/farmer ratio, but there was insufficient variability in the indicators to allow derivation of substantive conclusions. We thus opted to evaluate T & V extension, as a *system*, by comparing it with traditional extension.

COST-BENEFIT ANALYSIS

The final step in the evaluation required that the value of the increase in farm output, resulting from our estimates of the productivity differential attributable to T & V extension, be set against the additional costs incurred to make the additional output possible. To do so, the familiar technique of cost-benefit analysis was applied to the T & V system in Karnal (Feder *et al.*, 1985). Although the analysis was undertaken *ex post*, a complete series of either costs or benefits for the entire life of the project was not available. Therefore, we were obliged to make a number of assumptions redolent of an *ex ante* analysis.

The stream of incremental extension costs (adjusted to constant 1979 prices) was constructed using data on the actual costs of the first four years of T & V extension work in Karnal and projections made at the time of project appraisal. In scenarios where the project life was assumed to be less than the life span of physical structures and equipment, appropriate residual values were calculated and deducted from the costs.

As already mentioned, the increase in yield attributable to T & V extension was estimated for the third year of the project. However, it is reasonable to expect gains to continue to accrue over time. In the absence of data with which to estimate future (after the third year of the project) extension-induced gains, we were obliged to construct a dynamic model to simulate the future evolution of the change in productivity, both with and without T & V extension. The model is depicted graphically in Fig. 3.

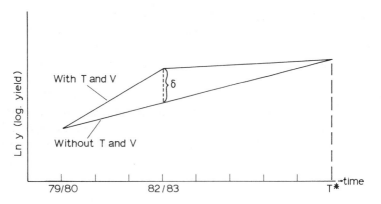

Fig. 3. Time pattern of productivity change, where δ is productivity gain due to extension in 1982/83.

This model assumes that in the absence of T & V extension, the average yield grows at a constant rate (estimated from a trend analysis of pre-project yields), and that after a few years the initial impetus to productivity resulting from the introduction of T & V slows down. Thus, after a number of years (T*), average yields, with and without T & V extension, are again the same. Therefore, the model conservatively assumes the supply of new knowledge from research to be limited. As a result, the gross benefits accruing from T & V extension are positive only in the years prior to T* in Fig. 3 — a period which can be called the 'project's life'. If it were possible to stop the project as soon as marginal benefits equal or are only slightly lower than marginal costs, a shorter project life would result, which may be called the 'efficient project life'. The model was used to estimate the project's total benefits for several variants of each of these concepts of project life. Finally, rather than calculate the corresponding internal rates of return, we chose to establish the level of statistical confidence with which a minimum acceptable rate of return could be expected.

MAIN EVALUATION RESULTS

The results of the evaluation showed that the diffusion of knowledge regarding improved farming practices for wheat was faster in the area covered by T & V extension work, but no clear superiority could be discerned for rice. Crop yields for farmers whose main source of information was the extension service were found to be higher than those of farmers who learned mainly from other sources.

After three years, the area where T & V extension was introduced had a 6–7 per cent gain in the productivity of wheat compared to the area with traditional extension. Significant results were not obtained for rice. The gains in wheat provide (at the 90 per cent confidence level) a rate of return of more than 15 per cent (more than 20 per cent using the concept of an efficient project life) to the incremental investment in T & V extension.

Most of the gains in yield induced by extension were found to derive from farmers' access (through the larger number of extension field workers) to timely and research-based advice concerning their specific production problems. The study area was very advanced even before the introduction of T & V extension, and no significant gain was identified in the use of high yielding varieties or fertilizers.

CONCLUSIONS

Although the methodology of the evaluation described above may have been adequate to cope with the particular circumstances that were confronted, a number of lessons seem clear.

We have several times alluded to weaknesses in our data base. Hence, if evaluation is to be efficient, suitable studies should begin long enough before the start of project implementation to collect sufficient material for a clear picture of the pre-project situation. Such work is not incompatible with normal pre-feasibility or preparation studies and indeed could be of material assistance. Data collection should continue at intervals until most adjustments resulting from the project intervention are complete. This may be long after the end of direct implementation. Without time series information spanning a sufficient period, it is difficult to obtain an understanding and a measure of the magnitude of the changes that have taken place. It goes almost without saying that the data collected should be of high quality. This requires painstaking attention to detail and a process of supervision which ensures that key variables are not omitted or imperfectly measured (see Slade and Feder, 1985).

Secondly, we conclude that the comparability of the subject area and the control area is crucial. The evaluation framework in Fig. 1 requires that the two areas be identical. In practice this is obviously unlikely, but it is, analytically, possible to control for small differences between them. These differences must, however, be minimized. By selecting geographically neighbouring areas, physical, climatic and cultural differences between the control and subject areas can be largely eliminated. If, however, as is the case with extension, the objective is to evaluate the effects on productivity of institutionally provided knowledge, which once released observes few natural or administrative boundaries, then the contamination of the control area by the subject area is possible. Minimizing this possibility requires the two areas to be separated by some feature (in our case a large river) which strongly inhibits such contamination.

These conclusions imply a third. Those who undertake these evaluation studies must approach their work with a clear idea of what they are trying to do. It matters little whether the aim is to evaluate the so-called direct changes or to measure the project's full impact. In either case, an explicitly defined evaluation scheme, a clearly articulated set of questions to be answered and a carefully constructed analytical framework must be in place before data collection begins. The asking of questions and the collection of data merely because they appear interesting will not yield a convincing evaluation, or advance our understanding of the processes of change, unless they are part of a systematic, consistent and analytical evaluation framework.

Discussion

The more those who manage and carry out extension activities understand the processes by which they are able to affect and change human behaviour, the greater is likely to be the effectiveness of their work. Such an understanding can be derived or improved if extension is supported by various forms of relevant research work as an integral part of the activities. Much of what is required is concerned with evaluating and monitoring extension work in relation to changes among its clients. However, there is also a need for more fundamental kinds of research. Since extension workers are primarily concerned with conveying information and advice and with initiating and supporting both individual and social action, it is necessary to investigate and understand human behaviour as well as how 'extension inputs' affect, i.e. are transformed and incorporated into, the activities of clients (such as their decision-making). There is also a need to be prepared for and to recognize unintended effects and to understand how they have come about. The discussions on research, evaluation and monitoring thus considered, from various points of view, the relations between theory (in the social and behavioural sciences) and appropriate applied forms of research, leading to models which could underpin this research and to relevant research methods.

In the discussion of research in general, it was recognized that extension research is a complex matter which can be considered at various levels. Three important areas were discussed. First, extension research can be a valuable adjunct to the management of extension services. If their managers recognize this, there is then a need to provide them with guidance on the allocation of the necessary resources to carry out relevant research. In practical terms, such research can be both of a 'routine' kind, which aims to help in maintaining or improving the effectiveness of an extension service, and of a 'special' variety which may occasionally be required (for example, if or when it is necessary to restructure an extension service).

Secondly, much of the recently developed understanding of knowledge systems provides important and appropriate models which can be applied in extension research. These can form appropriate frameworks of reference for relevant research on extension systems (and their elements) including their relationships to research institutions and to the farmer/client/user systems. Thirdly, much of the extension research which is required can be incorporated into farming systems research — in which the importance of including and involving the farmers at all stages of research is stressed. The involvement of farmers is particularly important in studies which aim to evaluate and monitor extension work in relation to its objectives.

The aims and outcomes of extension work, both material and non-material, are woven into a spiral of development so that what are objectives at one level or time become the means for achieving higher-order objectives. Thus, although it may be debatable whether particular objectives are of a material or of a non-material kind (a distinction which ideologically is of importance), this has little bearing on the processes of evaluating and monitoring them. For practical purposes, many behavioural objectives, such as group cohesiveness, or changes in knowledge, attitudes and skills, can be regarded as non-material since they cannot be given a direct financial value. Moreover, in the early stages of development, non-material objectives are particularly important in order to build a foundation for self-determining activity by the clients of extension workers. Such objectives would include, for example, the development of analytical skills and the improvement of bargaining abilities which, in turn, become means for rural people to relate effectively with external agencies which operate in an area. The evaluation of such non-material aspects can help to determine the stage in the development process which people have reached and, thereby, to monitor their progress. It is also important as 'formative' evaluation. The evaluation of past objectives can be used as a basis for deciding future communication and extension strategies. The evaluation of people's present knowledge, attitudes and practices can avoid the initiation of extension campaigns on what is already known. It can also strengthen a communication strategy by building upon existing knowledge, attitudes and values, and lead to particular attention being given to those factors which are identified as barriers to development.

Frequently, organizational objectives arise from plans of action derived from situation analyses made by 'specialists'. Thus, their focus will invariably reflect the specialists' interests, expectations, and perception of the problems. The need in extension work is to deal with people's (the clients') problems as they see them; in so far as extension work tries to solve a government's problems it may exacerbate and increase the number of problems for the people. Rural people, however, being accustomed to their local environment and social system, often find it very difficult to define their problems. One method, which has been used in rural Thailand, has sought to help people to do this through village workshops. At these, the villagers, initially in groups, have been asked to note their farming problems on large sheets of paper. These are then shown to everyone, and all are asked to rank the problems in order of importance and, for those which are dominant, to analyse what they consider to be the underlying reason(s) for the difficulties. Finally, they are invited to suggest how the problems might be solved using locally available or obtainable resources. This method of incorporating evaluation by the clients into the extension process has been found to accelerate the pace of local development provided it is undertaken at an appropriate time: sufficiently early to allow the results to be used in meetings concerned with deciding policy and the budget, but not so early that the time between the planning workshop and the beginning of extension activity becomes too long.

The processes of evaluation and monitoring depend critically on the choice of appropriate indicators. Preferably, these should be derived from the identification and descriptions of relevant variables being given by villagers, with appropriate indicators of them being based on joint decisions by the rural people and an expert. The criteria for a 'good' indicator would include: a particular variable could be assumed with a reasonable degree of accuracy by a single indicator (or, at most, by only a few); the required measurement would be relatively easy, quick, and made at low cost; and some numerical value could be given to the indicator.

Methods which are more sophisticated are normally required to provide economic assessments of the effects of extension work in terms of the rate of return on the investments in it. The ideal methodology for such research work would require an experimental design, but, in practice, this is rarely, if ever, possible. Therefore, comparisons have to be made between 'control' and 'treatment' areas. Economic evaluations can then be based on comparisons of changes in farmers' knowledge, in the operations of extension services, and in agricultural productivity, as well as in their interrelationships. The example which formed the basis for the discussion drew upon research work on evaluating the T & V system in India; this showed an internal rate of return of 15 per cent (at a 90 per cent level of confidence). This gave rise to many comments and queries concerning the longer-term implications of the results (such as, the effects on households' internal allocation of labour and the division of returns, and the role of the T & V system at different stages of agricultural development), and relating to the applicability of the methodology and the relevance of the results in other situations.

Postscript

The preceding papers and discussion summaries have considered many of what are the dominant features and issues of concern in rural extension today. They have done so from backgrounds of differing experiences and from various points of view. They have contributed to an integration of concepts drawn from several disciplines with practical experience in rural extension work, to narrowing and bridging the gap between theory and practice. Many have offered critical appraisals concerning the functions and value of extension activities and have raised important questions. This is timely, for the organization and conduct of extension work, internationally, appears to have reached a point of major decision regarding its future form and purpose.

Rural people, in Third World nations as much as in those which are developed, are currently undergoing significant social change. In both, however, there exist great varieties of rural cultural systems — ways of life as they are and as they have been, and the ways of what life-is-desired-to-be. Inextricably associated with these differing values and aspirations are the immense variations in their systems of production, primarily in their systems of farming and associated activities. The paths of change will be equally diverse. However, rural people are adjusting ever more rapidly towards new orders within the contexts of larger societal and technological changes which are themselves continuously altering. Even so, the pace at which the rural societies are changing and the ways in which they alter may often be regarded as insufficient or inappropriate by those who administer national and international development policies.

In order to be able to know of and be able to consider appropriate technological possibilities, people require relevant information and advice to help them to make rational judgements. So do they also in terms of different social and socio-economic possibilities in order to assess their cultural acceptability and likely impacts. This implies the existence of knowledge systems which not only convey adequate, reliable information but which also allow its recipients to influence its content and form in relation to their needs. It is in this sense that the extension work conducted among rural people, which can form a significant part of their modern knowledge system, linking them to a world of information external to their indigenous sources, is having to face a major reappraisal.

Even twenty years ago, it was generally accepted (as it had been for the previous century) that the main purpose of extension work in rural areas was to convey the findings of agricultural research and development, along with their economic implications, to the farm population. Over the past two decades or so, it has been recognized that a similar approach could be applied to many other facets of rural life (such as nutrition, health, and community affairs). Moreover, the many intricate interdependencies between agricultural and rural development have become more generally appreciated. It has also been realized, however, that the introduction and progressively wider diffusion of innovations, whether technological, economic, social or ideological, in the existing rural social systems are not peripheral or merely ineffectual appendages. Individually, and even more in combination, the innovations are among the central forces bringing about major changes in their institutions, social organizations, and individual patterns of behaviour. Modern systems of disseminating information (such as extension services) play a crucial role in introducing and establishing the acceptance of the innovations, and are thus responsible, in large measure, for the consequent social changes.

Possibly of even greater significance than these processes which generate rural change has been the growing acceptance of a different ethos underlying its planning and stimulation, particularly at a local level. Rural people certainly need to know and have an understanding of the technological possibilities and economic potentials open to them, as well as of the requirements of national policies, but the changes which they adopt should not be dictated solely by the apparent imperatives of modern technology or of national economy and policy. Rather, they should also be perceived by the people, individually and collectively, as being in their own best interests. To this end, they should be allowed and encouraged to participate actively in the processes which, at the same time, can both create in them a more overt consciousness of their problems and needs, and enable them to decide on the adjustments they should and could make in order to move towards achieving betterment in their lives.

Several contributions in this volume indicate that a good deal of experience of assisting rural people to become involved in such processes already exists among extension workers, especially those attached to non-governmental agencies or to special projects. The technocratic and paternalistic character, traditionally common in much agricultural and other rural extension work, is largely rejected in favour of helping the people, as individuals and as groups, to discover their real needs, to recognize the potentials which are available to them, and to work towards achieving their own aims within those of their larger societies. Work of this nature places a strong emphasis on the provision of sensitive, relevant guidance to stimulate the people concerned to develop their own resources (human as well as physical). It is an approach which has to be flexible, adapted to particular cultural situations.

For government extension services to be able to operate in this manner generally requires a difficult re-direction of the strategies, and even more of the policies, which underpin and authorize their work. This implies not so much changes in the content of their extension activities as the use of different methods and, especially, the adoption of alterations to their organization and management and in the criteria for their evaluation. It also involves different approaches to the training of extension personnel, to administering their deployment in rural areas, and to judging their performance in which they would necessarily have to be more accountable to their clients.

In developed countries, at least in the context of agricultural extension, this may be resolved in part by requiring the farmers to pay directly for their extension services; it might extend even to the privatization of extension organizations. This means that farmers explicitly recognize that information and advice is a necessary resource for the development of their businesses. Apart from each client's degree of satisfaction with the advisory 'inputs' derived from their extension services, this poses many problems for evaluating the more general technological, economic and social impacts of the work. In particular, it may mean that those who most need extension work deny themselves of the available (but costly) information, advice and guidance. And, it may be difficult to justify that the number of 'gainers' and the benefits they derive outweigh the number of 'losers'. In Third World countries, direct payment could probably be expected only from the commercial sector of agriculture (plantations, estates, and the minority of relatively large-scale farmers). But the need is to reach and assist the millions of small farmers and their families. In this, different and more flexible organizational arrangements are needed for conducting extension work, based on a policy which seeks to enhance the welfare and well-being of the rural poor. Welfarism, however, is rarely the explicit foundation for extension work; indeed, it might be considered culturally inappropriate in societies without western traditions of social welfare (or even a further imposition of alien values). The appropriate basis, in the developing world, is probably to be found in the social structures and organizations of their rural communities, with their own systems of providing mutual support. However, just as communities differ in their detailed compositions and structures, so there can be no one simple form of the approach which will serve for the entire spectrum of cultural difference which exists among them. Thus, any approach which is welfare-oriented must also be pluralistic. In its implementation, innovations in communications technology if used imaginatively can play an important role in enhancing the work's effectiveness (and there are many instances of this already occurring). To be able to assure that the rapidly growing rural population of the Third World has easy access to forms of trustworthy information and guidance which will enable them to improve their lives through the use and development of their own resources is the major challenge facing rural extension.

The theme of investing in rural extension is thus not only complex but it consists of many forms. At any level from the personal to the national, from the local to the societal, it is a composite of physical, technological, economic and social components. Moreover, in all cases, it requires a future orientation. In so far as extension workers have assisted individuals and communities to reach decisions, they have always been concerned with future states. But the present era of accelerating change imposes a need for all who devise and implement policies to think in more overtly futurological ways. In this, it is impossible to disregard past and present conditions and experiences. But more is involved than the extrapolation of trends, or even of reacting to the probable outcomes of these trends. A more distinctly prospective approach is neccesary, in which the futures which rural people desire become the bases for the kinds of knowledge systems which can assist them to move towards their expectations and aspirations. The desired futures are not so much goals as the means for devising appropriate policies and for adopting relevant processes which can lead the people towards realizing their hopes in ways which are equitable and satisfying. For, as they move, so will their horizons change. This will both assure that the need for rural extension continues and justify the investments made in it.

References and Bibliography

ADAB (Australian Development Assistance Bureau) (1981). *Final Report of the Thai-Australian Highland Agricultural Project to the Faculty of Agriculture, Chiang Mai University, and the Public Welfare Department of the Kingdom of Thailand.* Canberra: ADAB.

Adams, D. W. and Havens, A. E. (1966). The use of socio-economic research in developing a strategy for rural communication: A Columbian example, *Economic Development and Cultural Change,* **14,** 204–16.

Adams, M. E. (1982). *Agricultural Extension in Developing Countries.* London: Longmans.

Adelman, Irma and Morris, C. (1973). *Economic Growth and Social Equity in Developing Societies.* Stanford, California: Stanford University Press.

AERDC, University of Reading, and Extension Aids Branch, Ministry of Agriculture and Natural Resources, Malawi (1976). *Action Research and Media Production: a draft manual.* Reading: AERDC, University of Reading.

Albrecht, H. (1970). Application of socio-psychological research in extension education, *Sociologia Ruralis,* **10** (3), 237–52.

American Friends Service Committee (1984). Women and children last, *World Hunger Action Letter,* January/February. Washington, DC: American Friends Service Committee.

Anecksamphat, C. and Buddee, W. F. (1984). *A Critical Review of the Thai-Australian World Bank Project and Recommendations for the Future.* Chiang Mai (mimeo).

AOAP (The Australian Overseas Aid Program) (1984). *Report of the Committee to Review the Australian Overseas Aid Program.* Canberra: Australian Government Publishing Service.

Ascroft, Joseph R. (1974). A conspiracy of courtesy, *Intern. Development Review/ Focus,* no. 1, 8–11.

Ascroft, Joseph R., Röling, Niels, Kariuki, Joseph and Chege, Fred (1973). *Extension and the Forgotten Farmer.* Wageningen: Agricultural University, Social Science Department's Bulletin no. 37.

Ashok, Mehta (1978). *Report of the Committee on Panchayati Raj Institutions.* New Delhi: Government of India.

Ay, P. (1980). *Agrarpolitik in Nigeria. Produktionssystem der Bauern und die Hilfslosigkeit von Entwicklungsexperten.* Hamburg: Institut fur Agrikakunde.

Bell, Ronald L. (1984). *Report of a Study of ADAS by its Director General.* London: Ministry of Agriculture, Fisheries and Food.

Belloncle, Guy (1985). Proposals for a new approach to extension services in Black Africa. (Paper presented to the workshop on *Agricultural Extension and its Link with Research in Rural Development,* organized by the World Bank in Yamoussoukro, Ivory Coast).

Bennis, W. G., Benne, K. D. and Chin, R. (1961). *The Planning of Change: Readings in the Applied Behavioral Sciences.* New York: Holt, Rinehart and Winston.

Benor, D. and Baxter, M. (1984). *Training and Visit Extension.* Washington, DC: The World Bank.

Benor, D. and Harrison, J. Q. (1977). *Agricultural Extension: the Training and Visit System.* Washington, DC: The World Bank.

Benor, D., Harrison, J. Q. and Baxter, M. (1984). *Agricultural Extension: the Training and Visit System.* Washington, DC: The World Bank.

Berrigan, Frances (1981). *Community Communication: the Role of Community Media in Development.* Paris: UNESCO.

Bhaduri, A. and Rahman, M. A. (Editors) (1982). *Studies in Rural Participation.* New Delhi: Oxford and IBH Publishing Co.

Black, Cyril E. (1985). Quoted in: J. I. Merritt in *Campaign Bulletin* (Princeton University), Summer edition.

Bouman, F. (1984). Landbouwkrediet in ontwikkelingslanden, *Landbouwkundig Tijdschrift,* **96** (3), 23-8.

Boyce, James K. and Evenson, Robert E. (1975). *National and International Agricultural Research and Extension Programs.* New York: Agricultural Development Council.

Brakel, A. (Editor) (1985). *People and Organisations Interacting.* Chichester: John Wiley and Sons.

Brekelbaum, T. (1984). The use of paraprofessionals in rural development, *Community Dev. Journ.,* **19** (4), 232-45.

Bright, Simon (1981). *Media for Dialogue in Development.* AERDC, University of Reading, unpublished MA dissertation.

Bunting, A. H. (1979). Serving the rural poor: the role of science and technology, *Reading Rural Development Communications Bulletin* (AERDC), no. 6: 6-11.

Byram, Martin and Garforth, Chris (1979). *Iketleetse: an Experimental Multi-Media Rural Extension Project in Kgalagadi District, Botswana.* Gabarone: Agricultural Information Service and Matsha Community College.

Carruthers, Ian and Chambers, Robert (1981). Rapid appraisal for rural development, *Agric. Administration,* **8** (6), 407-22.

Cartwright, D. (Editor) (1951). *Field Theory in Social Science: selected theoretical papers of Kurt Lewin.* New York: Harper and Row.

Cernea, Michael M. (1983). *A Social Methodology for Community Participation in Local Investments: the experience of Mexico's PIDER program.* Staff Working Paper no. 598. Washington DC: The World Bank.

Cernea, Michael M. (1984*a*). Putting People First — the position of sociological knowledge in planned rural development. (Keynote opening address to the *VIth World Congress for Rural Sociology,* Manila, December, 1984.)

– Cernea, Michael M. (1984*b*). The role of sociological knowledge in planned rural development, *Sociologia Ruralis,* **24** (3–4), 185–201.

⌐ Cernea, Michael M. (Editor) (1984*c*). *Putting People First: Sociological Variables in Rural Development.* Baltimore: The Johns Hopkins University Press for the World Bank.

Cernea, Michael M., Coulter, John K. and Russell, John F. A. (Editors) (1983). *Agricultural Extension by Training and Visit: the Asian Experience.* (Proceedings of a World Bank and UNDP Symposium held in Chiang Mai, Thailand, December, 1981). Washington, DC: The World Bank.

Cernea, Michael M., Coulter, John K. and Russell, John F. A. (Editors) (1985). *Research-Extension-Farmer: A Two-way Continuum for Agricultural Development.* (Proceedings of a World Bank and UNDP Symposium held in Denpasar, Indonesia, March 1984.) Washington, DC: The World Bank.

Cernea, Michael M. and Guggenheim, Scott E. (1985). 'Is anthropology superfluous in farming systems research?' In: Flora and Tomecek (Editors).

Cernea, Michael M. and Tepping, B. J. (1977). *A System for Monitoring and Evaluating Agricultural Extension Projects.* Staff Working Paper no. 272. Washington, DC: The World Bank.

Cesarini, G. (1982). Serving small farms in marginal areas. Why and how? Problems involved in training for extension in agriculture. *Agric. Administration,* **11** (1), 39–48.

Chambers, Robert (1980*a*). The small farmer is a professional, *Ceres,* **13** (2), 19–23.

Chambers, Robert (1980*b*). *Rural Poverty Unperceived: Problems and Remedies.* Staff Working Paper no. 400. Washington, DC: The World Bank.

Chambers, Robert (1981). Rapid rural appraisal: rationale and repertoire, *Public Administration and Development,* **1**, 95–106. (See also the issue of *Agric. Administration,* **8** (6)).

Chambers, Robert (1983). *Rural Development: Putting the Last First.* London: Longmans.

Chambers, Robert and Ghildyal, B. P. (1985). Agricultural research for resource-poor farmers: the farmer-first-and-last model, *Agric. Administration,* **20** (1), 1–30.

Chambers, Robert and Wickremanayake, M. (1977). Agricultural extension: myth, reality and challenge, pp. 155–67 in: Farmer (Editor).

Chin, Robert (1961). The utility of system models and development models for practitioners, pp. 201–14 in: Bennis, Benne, and Chin (Editors).

China, Richard and Langmead, Peter (1985). Using videos to increase farm output, *International Agricultural Development,* **5** (3), 18–19.

Coletta, Nat J. (1979). Popular participation: the Sarvodaya experience. *International Development Review,* **15** (3), 15–18.

Colin, R. (1978). Analyse comparative et problematique a partir de l'etude de quelques dossier significatifs, in: *Les Methodes et Techniques de la Participation en Developpement.* Paris: UNESCO, Division for the Study of Development Projects.

Collinson, Michael (1981). A low cost approach to understanding small farmers, *Agric. Administration,* **8** (6), 433–50.

Collinson, Michael (1983). *Farm Management in Peasant Agriculture*. Boulder, Colorado: Westview Press.

Collinson, Michael (1984). On farm research with a systems perspective as a link between farmers, technical research and extension. (Paper presented at *Workshop on Extension and Research*, Eldoret, Kenya, organized by the World Bank.)

Commission of the European Communities (1984). *Women and Development*. Brussels: The Commission (X/297/84-EN).

Coombs, P. H. (Editor) (1980). *Meeting the Basic Needs of the Rural Poor. The Integrated Community-based Approach*. New York: Pergamon Press for the International Council for Educational Development.

Coombs, P. H. and Ahmed, M. (1974). *Attacking Rural Poverty: How Nonformal Education Can Help*. Baltimore: Johns Hopkins University Press.

Crouch, B. R. and Chamala, S. (Editors) (1981a). *Extension Education and Rural Development*. 2 vols. Chichester: John Wiley and Sons.

Crouch, B. R. and Chamala, S. (1981b). Communication strategies for technological change in agriculture: implications for rural society, pp. 257–313 in: Vol. 2, Crouch and Chamala (Editors).

De Datta, S. K., Gomez, K. A., Herdt, R. W. and Barker, R. (1978). *A Handbook on the Methodology for an Integrated Experiment Survey on Rice Yield Constraints*. Los Baños, The Philippines: International Rice Research Institute.

de Kadt, E. and Williams, G. (Editors) (1976). *Sociology and Development*. London: Tavistock Publications.

Devitt, Paul (1982). *The Management of Communal Grazing in Botswana*. London: Overseas Development Institute.

Dexter, K. (1984). Les services de vulgarisation, les firmes d'amont et d'aval, et les organisme de commercialisation, *Economic Rurale*, **159**, 67–71.

Dube, S. C. (1981). Crisis and commitment challenges to intellectual craftsmanship in the social sciences, *Regional Institute of Higher Education and Development Bulletin* (Singapore), **9** (3), 4–7.

Eberts, Paul M. and Kelly, Janet M. (1983). How mayors get things done: community politics and mayors' initiations. *Rural Sociology Bulletin Series*, no. 133. Ithaca, New York: Cornell University.

Elliot, Charles (1977). *Patterns of Poverty in the Third World*. New York: Praeger.

Esman, Milton J. (1983). *Paraprofessionals in Rural Development: Issues in Field Level Staffing for Agricultural Projects*. Staff Working Paper no. 573. Washington, DC: The World Bank.

Esman, Milton J. and Uphoff, Norman T. (1984). *Local Organizations: Intermediaries in Rural Development*. Ithaca, New York: Cornell University Press.

Esman, Milton J., Colle, R., Uphoff, N. and Taylor, E. (1980). *Paraprofessionals in Rural Development*. Ithaca, New York: Cornell University, Center for International Studies, Rural Development Committee.

Evenson, R. E. (1982). Agriculture, in: Nelson (Editor).

Evenson, R. E. (1984). Rural development experience: economic perspectives. (Keynote paper to the *VIth World Congress for Rural Sociology*, Manila, December, 1984.)

Evenson, R. E. (1985). *IARC Investment, National Research and Extension*

Investment, and Field Crop Productivity. New Haven, Conn.: Yale University (mimeo).

Evenson, R. E. and Jha, D. (1973). The contribution of the agricultural research system to agricultural production in India, *Indian Journ. Agric. Econ.,* **28** (4), 212–30.

Fals Borda, O. (1984). Participatory action research, *Journ. Society Internat. Dev.,* **2**, 18–20.

FAO (Food and Agriculture Organization of the United Nations) (1978). *Small Farmers Development.* 2 vols. Bangkok: FAO Regional Office for Asia and the Far East.

FAO (Food and Agriculture Organization of the United Nations) (1979). *World Conference on Agrarian Reform and Rural Development: Report.* Rome: FAO.

FAO (Food and Agriculture Organization of the United Nations) (1983). *FAO Small Farmers Development Programme in Asia and the Pacific.* Bangkok: FAO Regional Office for Asia and the Far East (mimeo).

Farmer, B. H. (Editor) (1977). *Green Revolution? Technology and Change in Rice Growing Areas of Tamil Nadu and Sri Lanka.* London: Macmillan & Co.

Feder, G. and Slade, R. H. (1984). Contact farmer selection and extension visits: the Training and Visit extension system in Haryana, India, *Quarterly Journ. Internat. Agric.,* **23** (1), 6–21.

Feder, G. and Slade, R. H. (1986). A comparative analysis of some aspects of the Training and Visit system of agricultural extension in India, *Journ. Dev. Studies,* **22** (2).

Feder, G., Lau, L. J. and Slade R. H. (1985). *The Impact of Agricultural Extension: A Case Study of the Training and Visit System (T & V) in Haryana, India.* Staff Working Paper no. 756. Washington, DC: The World Bank.

Fernandes, W. and Tandon, R. (Editors) (1981). *Participatory Research and Evaluation: Experiments in Research as a Process of Liberation.* New Delhi: Indian Social Institute.

Fewster, W. Jean and Kuhonta, C. M. (1985). Effective communication, *Canadian Home Economics Journ.,* **35** (2).

Flora, C. Butter and Tomecek, M. (Editors) (1985). *Selected Proceedings of 1984 Farming Systems Research Symposium.* Manhattan, Kansas: Kansas State University Research Series, Vol. 9.

Foster, P. and Sheffield, R. J. (Editors) (1973). *Education and Rural Development.* London: Evans Bros.

Fraser, Colin (1980). Video in the field. *Ceres,* **13** (1), 24–28.

FSSP (Farming Systems Support Project) (1985). *Training Manuals and Materials for Farming Systems Research and Extension.* Gainesville, Florida: University of Florida.

Fuglesang, A. (1979). *Applied Communication in Developing Countries. Ideas and Observations.* Uppsala: The Dag Hammarskjold Foundation.

Fyson, N. (1983). *A Woman's Place. . . .* London: Centre for World Development Education.

Garforth, Chris (1982*a*). Botswana: action research in media production, *Reading Rural Development Communications Bulletin* (AERDC) no. 16: 16–18.

Garforth, Chris (1982*b*). Reaching the Rural Poor: a review of extension strategies and methods, pp. 43–70 in: Jones and Rolls (Editors).

Garforth, Chris (1983). Who do you think you're talking to? *Media in Education and Development,* **16** (1), 9–12.

Gilles, J. L. and Jamtgaard, K. (1981). Overgrazing in pastoral areas: the commons reconsidered, *Sociologia Ruralis,* **21** (2), 129–41.

Giltrow, D. and Potts, J. (1978). *Agricultural Communication: The Role of the Media in Extension Training.* London: The British Council.

Gittinger, J. P. (1982). *Economic Analysis of Agricultural Projects.* Second Edition. Baltimore: The Johns Hopkins University Press and IBRD/The World Bank.

Gotsch, C. H. (1972). Technical change and the distribution of income in rural areas, *Amer. Journ. Agric. Econ.,* **54** (2), 326–41.

Griliches, Z. (1958). Research costs and social returns: hybrid corn and related innovations, *Journ. Pol. Economy,* **66**, 419–31.

Grosser, E. and Ibra Ba, A. (Editors) (1979/80). *Analyse de Situation de la Region du Tagant (Republique Islamique de Mauretanie) avec attention particuliere aux aspects socio-economique.* Berlin: Technische Universität, Institut fur Sozioökonomie der Agrarentwicklung. (Reihe Studien IV/26).

Halim, A. (1971). The economic contribution of schooling and extension to rice production in Laguna, Philippines, *Agric. Economics and Development,* **7**, 33–6.

Hall, B. L. (1977). *Notes on the Development of the Concept of Participatory Research in an International Context.* Toronto: International Council for Adult Education.

Haque, W., Mehta, N., Rahman, A. and Wignaraja, P. (1977). Towards a theory of rural development, *Development Dialogue,* no. 2. 11–137.

Hardin, Garrett (1968). The tragedy of the commons, *Science,* **162**, 1243–48.

Hardin, Garrett and Baden, John (Editors) (1977). *Managing the Commons.* San Francisco: W. H. Freeman & Co.

Harker, B. R. (1973). The contribution of schooling to agricultural modernization: an empirical analysis, pp. 350–71 in: Foster and Sheffield (Editors).

Havelock, R. C. and collaborators (1969). *Planning for Innovation through Dissemination and Utilization of Knowledge.* Ann Arbor, Michigan; CRUSK, Institute for Social Research, The University of Michigan.

Haverkort, Bertus and Röling, Niels (1984). *Six Approaches to Rural Extension.* Wageningen: International Agriculture Centre.

Heaver, Richard (1982). *Bureaucratic Politics and Incentives in the Management of Rural Development.* Staff Working Paper no. 537. Washington, DC: The World Bank.

Heaver, Richard (1984). *Adapting the Training and Visit System for Family Planning, Health and Nutrition Programs.* Staff Working Paper no. 662. Washington, DC: The World Bank.

Hildebrand, Peter (1981). Combining disciplines in rapid appraisal: the Sondeo approach, *Agric. Administration,* **8** (6), 423–32.

Hitchcock, Robert K. (1980). Tradition, social justice and land reform in Central Botswana, *Journ. African Law,* **24** (1), 1–34.

Hoare, P. W. C. (1984*a*). The declining productivity of traditional highland farming systems in Northern Thailand, *Thai Journ. Agric. Sci.,* **17** (2).

Hoare, P. W. C. (1984*b*). Improving the effectiveness of agricultural development:

a case from North Thailand, *Manchester Papers on Development*, no. 10, 13–43.

Hoare, P. W. C. (1985). Farmer-centered extension to increase adoption of new technology: experience in Northern Thailand, pp. 154–161 in: Cernea, Coulter and Russell (Editors).

Hoare, P. W. C. and Wells, G. J. (1984). The adoption of improved cultural practices for paddy rice in North Thailand. (Paper presented at the *VIth World Congress for Rural Sociology,* Manila, December, 1984.)

Hoare, P. W. C., Crouch, B. R. and Lamrock, J. C. (1982). Methodology of rural development in the uplands and highlands of North Thailand, *Thai Journ. Agric. Sci.,* **15** (4).

Hoeper, B. (1983). *Selected Results of ADO and VEW Survey in Jind, Karnal, and Mahendragarh Districts of Haryana, India.* Berlin: Institute of Socio-Economics and Agricultural Development (Working Note no. 1) (mimeo).

Houseman, Ian (1981). Viewdata and Teletext — tomorrow's information today, *Agricultural Progress,* **56,** 91–6.

Howell, J. (1982*a*). *Managing Agricultural Extension: the T & V system in practice.* London: Overseas Development Institute, Agricultural Administration Unit (Discussion Paper no. 8).

Howell, J. (1982*b*). Managing agricultural extension: the T & V system in practice, *Agric. Administration,* **11** (4), 273–84.

Howell, J. (1984). *Conditions for the Design and Management of Agricultural Extension.* London: Overseas Development Institute, Agricultural Administration Unit (Discussion Paper no. 13).

Howes, M. and Chambers, Robert (1978). Indigenous technical knowledge: analysis, implications and issues, *IDS Bulletin,* **10** (2), 5–11.

Hudson, Heather E. (1984). *When Telephones Reach the Village: the Role of Telecommunication in Rural Development.* Norwood, New Jersey: Ablex Publishing Corporation.

Huffman, W. E. (1974). Decision-making: the role of education, *Amer. Journ. Agric. Econ.,* **56** (1), 85–97.

Huffman, W. E. (1976*a*). The productive value of human time in US agriculture, *Amer. Journ. Agric. Econ.,* **58** (4), 672–83.

Huffman, W. E. (1976*b*). The value of the productive time of farm wives: Iowa, North Carolina, and Oklahoma, *Amer. Journ. Agric. Econ.,* **58** (5), 836–41.

Huffman, W. E. (1977). Allocative efficiency: the role of human capital, *Quarterly Journ. Economics,* **91,** 59–79.

Huffman, W. E. and Miranowski, J. A. (1981). An economic analysis of expenditures on Agricultural Experiment Station research, *Amer. Journ. Agric. Econ.,* **63** (1), 104–18.

Huizer, G. (1982). *Guiding Principles for People's Participation Projects.* Rome: FAO.

Hunter, Guy (Editor) (1978*a*). *Agricultural Development and the Rural Poor.* London: Overseas Development Institute.

Hunter, Guy (1978*b*). Small farmers and rural employment; signs of a new total approach in India and Pakistan, *Agric. Administration,* **5** (2), 111–20.

Hunter, Guy (Editor) (1982). *Enlisting the Small Farmer: The Range of Requirements.* London: Overseas Development Institute, Agricultural Administration Unit, Occasional Paper no. 4.

Hunter, Guy and Jiggins, Janice (1977). *Farmer and Community Groups*. London: Overseas Development Institute, Agricultural Administration Unit (mimeo).

Hyden, G. (1983). *No Shortcuts to Progress: African Development Management in Perspective*. London: Heinemann.

IIC (International Institute of Communications) (1983). India: using video for local change, *Intermedia*, **11** (3).

Illich, Ivan (1969). Outwitting the 'developed' countries, *New York Review of Books*, 6th edition (November).

Jamison, Dean T. and Lau, Lawrence, J. (1982). *Farmer Education and Farm Efficiency*. Baltimore: The Johns Hopkins University Press for the World Bank.

Jiggins, J. (1983*a*). Poverty-oriented rural development: participation and management, *Development Policy Review*, **1**, 219–52.

Jiggins, J. (1983*b*). *Rural Women: Their Role and Functions*. Wageningen: International Agricultural Centre.

Jones, G. E. and Rolls, M. J. (Editors) (1982). *Progress in Rural Extension and Community Development*, Vol. 1. Chichester: John Wiley and Sons.

Judd, M. Ann, Boyce, James K. and Evenson, Robert E. (1983). Investing in Agricultural Supply. New Haven: Yale University, Economic Growth Centre (Discussion Paper 442).

Karnik, Kiran (1981). Developmental television in India, *Educational Broadcasting International*, **14** (3), 131–35.

Kidd, R. (1982). Botswana, Nigeria: participatory drama, popular analysis and conscientizing the development worker, *Reading Rural Development Communications Bulletin*, (AERDC), no. 16, 19–25.

Kitching, G. (1982). *Development and Underdevelopment in Historical Perspective*. London: Methuen.

Korten, David (1980). Community organization and rural development: a learning process approach, *Public Admin. Rev.*, **40** (5), 480–511.

Korten, Francis (1982). *Building National Capacity to Develop Water Users' Association: Experience from the Philippines*. Staff Working Paper no. 528. Washington, DC: The World Bank.

Korten, Francis and Young, Sarah (1978). The Mothers Club of Kuala Lumpur, in: *Managing Community-Based Population Programmes*. Kuala Lumpur: International Committee for the Management of Population Programmes.

Laflin, Mike (1982). Alaska, Peru: the non-distance-sensitive satellite, *Reading Rural Development Communications Bulletin*, no. 16, 26–9.

Lamrock, J. C. (1966). Problem census technique, in: *Extension Manual*. Port Moresby, P. N. G.: Department of Agriculture, Stock and Fisheries.

Lamrock, J. C. (1979). *Report to the Australian Development Bureau on Extension Strategies for Highland Development, North Thailand*. St. Lucia: University of Queensland.

Lappé, Francis Moore, Collins, Joseph and Kinley, David (1980). *Aid as Obstacle: Twenty Questions about our Foreign Aid and the Hungry*. San Francisco: Institute for Food and Development Policy.

Lele, U. (1975). *The Design of Rural Development: Lessons from Africa*. Baltimore: The Johns Hopkins University Press.

Leonard, D. (1973). Why do Kenya's agricultural extension services favor the rich farmer? (Paper read at the *16th Annual Meeting of the African Studies Association,* Syracuse, New York.)

Leonard, D. (1977). *Reaching the Peasant Farmer: Organisation Theory and Practice in Kenya.* Chicago: Chicago University Press.

Leonard, D. K., Cohen, J. M. and Pinckney, T. C. (1983). Budgeting and financial management in Kenya's agricultural ministries, *Agric. Administration,* **14** (2), 105–20.

Lionberger, Herbert F. (1982). Toward an idealized system model for generating and utilizing information for a modernizing agriculture: a third attempt. (Paper presented to the *Conference on Knowledge Utilization: Theory and Methodology,* East-West Center, Honolulu.)

Lockheed, M. E., Jamison, D. T. and Lau, L. J. (1980). Farmer education and farm efficiency: a survey, *Economic Development and Cultural Change,* **29** (1), 38–76.

Maalouf, W. D. and Contado, T. E. (1984). Basic and in-service training for agricultural extension, pp. 39–47 in: *1983 Training for Agriculture and Rural Development.* Rome: FAO (Economic and Social Development Series no. 31).

McAnany, E. G. (Editor) (1980). *Communications in the Rural Third World.* New York: Praeger.

McCabe, Maria S. and Swanson, B. E. (1975). *International Directory of Extension Organizations and Extension Training Institutions.* Champaign, Ill: University of Illinois at Urbana Champaign.

McGubbin, Stephen (1985). Project SHARE, *Media in Education and Development,* **18** (1), 42–4.

McNamara, Sister Nora (1983). A plea for the inclusion of the Socio-Cultural in the process of Ecclesial and Institutional Planning and Organization. (Unpublished paper presented to World Bank staff.)

Mann, Charles K. (1978). Packages of practices: a step at a time with clusters, *Studies in Development* (Middle East Technical University, Ankara), **21**, 73–82.

Mazumdar, V. (Editor). (1978). *Role of Rural Women in Development.* Bombay: Indian Council of Social Science Research.

Mermillod, M. J. (1984). International programmes for the development of rural women, *Home Economics* (International Federation of Home Economics), **244/245** (3-4), 10–12.

Mohan, R. and Evenson, R. E. (1975). The Intensive Agricultural District Program in India: a new evaluation, *Journ. Dev. Studies,* **11**, 135, 154.

Mooch, P. R. (1976). The efficiency of women as farm managers: Kenya, *Amer. Journ. Agric. Econ.,* **58** (5), 831–5.

Mooch, P. R. (1978). Education and Technical Efficiency in Small Farm Production. Columbia University, unpublished manuscript.

Moore, M. (1984). Institutional development, the World Bank, and India's new agricultural extension programme, *Journ. Dev. Studies,* **20** (4), 303–17.

Moris, Jon R. (1981). *Managing Induced Rural Development.* Bloomington, Indiana: International Development Institute.

Moris, Jon R. (1983). *Reforming Agricultural Extension and Research Services in Africa.* Discussion Paper no. 11. London: Overseas Development Institute, Agricultural Administration Unit.

Mosher, A. T. (1978). *An Introduction to Agricultural Extension.* New York:

Agricultural Development Council.

Nelson, R. R. (Editor) (1982). *Government and Technical Progress.* Elmsford, New York: Pergamon Press.

Nitsch, U. (1982). *Farmers' Perceptions of, and Preferences Concerning Agricultural Extension Programs.* Uppsala: Institute for Economics and Statistics, Swedish University of Agricultural Sciences, Report no. 195.

Oakley, P. (1984). *The Monitoring and Evaluation of Participation in Rural Development.* Rome: FAO.

Oakley, P. and Marsden, D. (1984). *Approaches to Participation in Rural Development.* Geneva: ILO.

Oakley, P. and Winder, D. (1981). The concept and practice of rural social development: current trends in Latin America and India, *Manchester Papers on Development,* no. 1: 1-71.

Odell jr, Malcolm J. (1980). *Sociological Research and Rural Development Policy.* Gabarone: Government Printer (Ministry of Agriculture, Rural Sociology Report Series no. 19).

Odell jr, Malcolm J. (1982). *Local Institutions and Management of Communal Resources: Lessons from Africa and Asia.* London: Overseas Development Institute.

Odell jr, Malcolm J. and Odell, Marcia (1980). *The Evolution of a Strategy for Livestock Development in the Communal Areas of Botswana.* London: Overseas Development Institute.

Odell, Marcia L. (1979). *Divide and Conquer: Allotment among the Cherokee.* New York: Arno Press/New York Times.

Opio-Odongo, J. M. A. (1985). Factors complicating interdisciplinary collaboration in rural development, *The Rural Sociologist,* **5** (4), 230-7.

Orivel, F. (1983). The impact of agricultural extension: a review of the literature, pp. 1-58 in: Perraton, Jamison, Jenkins, Orivel, and Wolff (Editors).

Oxby, Clare (1982). *Group Ranches in Africa.* London: Overseas Development Institute.

Oxby, Clare (1983). 'Farmer groups' in rural areas of the Third World, *Community Development Journ.,* **18** (1), 50-9.

Paddock, W. and Paddock, E. (1973). *We Don't Know How.* Ames, Iowa: Iowa State University Press.

Parish, E. (1948). *Crop Survey of the Bechuanaland Protectorate.* Botswana National Archive, Box 500, no. S. 500/20.

Patrick, G. F. and Kehrberg, E. W. (1973). Costs and returns of education in five agricultural areas of Eastern Brazil. *Amer. Journ. Agric. Econ.,* **55** (2), 145-53.

Pearse, A. and Stiefel, M. (1979). *Enquiry into Participation.* Geneva: UNRISD.

Perraton, H., Jamison, D. T., Jenkins, Janet, Orivel, F. and Wolff, L. (Editors) (1983a). *Basic Education and Agricultural Extension: Costs, Effects, and Alternatives.* Staff Working Paper no. 564. Washington, DC: The World Bank.

Perraton, Hilary, Jamison, Dean, and Orivel, Francois (1983b). Mass media for agricultural extension in Malawi, pp. 147-201 in: Perraton, Jamison, Jenkins, Orivel, and Wolff (Editors).

Peters, J. A. (1985). Reaching Rural Women (Paper presented to *FAO Expert Consultation on Rural Youth and Young Farmers in Developing Countries,*

Rome).

Peterson, W. L. (1969). The allocation of research, teaching and extension personnel in US Colleges of Agriculture, *Amer. Journ. Agric. Econ.*, **51** (1), 41–56.

Pitt, David (1976). *The Social Dynamics of Development.* Oxford: Pergamon.

Planning, Research and Action Institute (1963). *Action Research and its Importance in an Underdeveloped Economy.* Lucknow: Planning, Research and Action Institute.

Potts, M. J. (Editor) (1983). *On-farm Potato Research in the Philippines.* (A collection of reports by staff of the International Potato Center and the Philippines Council for Agriculture and Resources Research and Development.)

Proceedings of the Workshop on Agricultural Extension and its Link with Research in Rural Development, Yamoussoukro, Ivory Coast, February, 1985. Washington, DC: The World Bank (forthcoming).

Rahim, S. A. (1966). The Comilla Program in East Pakistan, pp. 415–24 in: Wharton (Editor).

Rahman, M. A. (1982). The theory and practice of participatory action research, *Dossier* (International Foundation for Development Alternatives, Nyon, Switzerland), no. 31, 17–29.

Raintree, J. B. (1984). A systems approach to agro-forestry diagnosis and design: ICRAF's experience with an interdisciplinary methodology. (Paper presented at the *VIth World Congress for Rural Sociology,* Manila, December, 1984.)

Rhoades, Robert E. and Booth, Robert H. (1982). Farmer-back-to-farmer: a model for generating acceptable agricultural technology, *Agric. Administration,* **11** (2), 127–37.

Rogers, Everett, M. (Editor) (1976). *Communication and Development: Critical Perspectives.* Beverly Hills and London: SAGE Publishers.

Rogers, Everett M. (1983). *The Diffusion of Innovations.* Third Edition. New York: The Free Press.

Rogers, Everett M., Eveland, J. D. and Bean, A. S. (1976). *Extending the Agricultural Extension Model.* Stanford, California: Stanford University, Institute for Communication Research.

Röling, N. (1974). Problem-solving research. A strategy for change, *Tijdschrift voor Agologie,* **2**, 66–73.

Röling, N. (1982). Alternative approaches to extension, pp. 87–115 in: Jones and Rolls (Editors).

Röling, N. and de Zeeuw, H. (1983). *Improving the Quality of Rural Poverty Alleviation.* (Final Report of the Working Party on 'The Small Farmer and Development Cooperation'.) Wageningen: International Agricultural Centre.

Röling, N., Ascroft, J. and Chege, F. W. (1976). The diffusion of innovations and the issue of equity in rural development, *Communication Research,* **3** (2), 155–70.

Rondinelli, Dennis A. (1982). The dilemma of development administration: complexity and uncertainty in control-oriented bureaucracies, *World Politics,* **35** (1), 43–72.

Rose-Ackerman, Susan and Evenson, Robert E. (1985). The political economy of agricultural research and extension: grants, votes, and reapportionment, *Amer. Journ. Agric. Econ.,* **67** (1), 1–14.

Roy, A. (1985). Appropriate training for women's development: Another trap? (Paper presented to *FAO Expert Consultation on Rural Youth and Young Farmers in Developing Countries,* Rome.)

Runge, Carlisle F. (1983). Common property and collective action in economic development. (Paper for Board on Science and Technology Development.) Washington, DC: National Research Council.

Rural Women (1984). Special issue of *Ideas and Action,* no. 158. Rome: FAO (Freedom from Hunger/Action for Development).

Russell, John F. A. (1985). Essential ingredients of an effective extension service and some issues arising from World Bank experience in Sub-Saharan Africa, in: *Proceedings of the Annual Research and Extension Programme Review Conference.* Harare: Ministry of Agriculture (Research and Special Services Agency, and AGRITEX).

Sandford, Stephen (1980). *Keeping an Eye on TGLP.* Gabarone: University College of Botswana.

Sanusi, N. A. (1983). *Agricultural Development Policies in the Third Five Year Development Plan in Indonesia.* Jakarta: Ministry of Agriculture, Agency for Agricultural Education, Training and Extension.

Saunders, S. and Smith, W. A. (1984). Social marketing: two views, two opportunities, *Development Communications Report.* (Clearinghouse on Development Communication, Washington, DC), no. 47 (Autumn): 1–2.

Savile, A. H. (1965). *Extension in Rural Communities.* London: Oxford University Press.

Scoullar, B. B. (1984). The development of extension services: a result and a methodology. (Paper presented to the *VIth World Congress for Rural Sociology,* Manila, December, 1984.)

Searle, J. (1984). *Minds, Brains and Science.* (The 1984 Reith Lectures.) London: British Broadcasting Corporation.

Shaner, W. W., Philipp, P. F. and Schemehl, W. R. (1982). *Farming Systems Research and Development: Guidelines for Developing Countries.* Boulder, Colorado: Westview Press.

Shingi, Prakash M. and Mody, Bella (1976). The communication effects gap: a field experiment on television and agricultural ignorance in India, pp. 79–99 in: Rogers (Editor).

Simmonds, N. W. (1985). *The State-of-the-Art of Farming Systems Research.* Washington, DC: The World Bank (Agriculture and Rural Development Department, Technical Paper 43).

Simpson, Rowton (1967). The new land law in Malawi, *Journ. Admin. Overseas,* **6** (4).

Slade, R. H. and Feder, G. (1985). *The Monitoring and Evaluation of Training and Visit Extension: A Manual of Instruction.* Washington, DC: The World Bank (mimeo).

Smith, Cameron L. and Tippett, Bruce A. (1982). *Study of Problems Related to Scaling-up Micro-Enterprise Assistance Programs. Phase I.* Washington, DC: The World Bank.

The State of the World's Women (1985). *An information pack for the World Conference to Review and Assess the Achievements of the International Decade for Women, Equality and Peace.* Oxford: New Internationalist Publications Corp.

Stocking, Michael and Abel, Nick (1981). Ecological and environmental indicators for the rapid appraisal of natural resources, *Agric. Administration,* **8** (6), 473–84.

Sukaryo, D. G. (1983). Farmer participation in the Training and Visit System and the role of the village extension worker; experience in Indonesia, pp. 17–25 in: Cernea, Coulter and Russell (Editors).

Swanson, B. (Editor) (1984). *Agricultural Extension: A Reference Manual.* Second Edition. Rome: FAO.

Tully, J. (1981). Changing practices: A case study, pp. 79–85 in: Crouch and Chamala (Editors), Vol. 2.

Turnbull, C. M. (1963). *The Lonely African.* New York: Anchor Books.

Ugboajah, F. O. (1972). Traditional-urban media model: stocktaking for African development, *Gazette,* **18,** 76–95.

UNDP (1980). *Rural Women's Participation in Development.* (Evaluation Study no. 3.) New York: UNDP.

United Nations (1984). *Guiding Principles for the Design and Use of Monitoring and Evaluation in Rural Development Projects and Programmes.* Rome: The United Nations ACC Task Force on Rural Development (Panel on Monitoring and Evaluation).

Uphoff, Norman T. (1984). Participation in development initiatives: fitting projects to people, in: Cernea (Editor).

Uphoff, N. T., Cohen, J. and Goldsmith, A. D. (1979). *Feasibility and Application of Rural Development Participation: A state-of-the-art paper.* Ithaca, New York: Cornell University, Rural Development Committee (Monograph Series).

USAID (United States Agency for International Development) (1978). *The Dam at Laur* (film produced by S. J. Staniski for Training Division). Washington, DC: USAID.

van de Laar, A. (1982). *The World Bank and the Poor.* The Hague: Kluwer-Nijhof Publishing.

van den Ban, A. W. (1982). *Introduction to Extension Education.* Sixth Edition. (In Dutch.) Meppal: Boom (An English version is in preparation in collaboration with H. S. Hawkins).

Verhagen, K. (1984). *Cooperation for Survival: Analysis of an Experiment in Participatory Research and Planning with Small Farmers in Sri Lanka and Thailand.* Amsterdam: Royal Tropical Institute.

von Blanckenburg, P. (1982). The training and visit system in agricultural extension. A review of first experiences, *Quarterly Journ. of Intern. Agric.,* **21** (1), 6–25.

von Blanckenburg, P. (1984). *Agricultural Extension Systems in Some Asian and African Countries.* Rome: Food and Agricultural Organization of the United Nations.

von Fürer-Haimendorf, C. (1964). *The Sherpas of Nepal.* Berkeley and Los Angeles: University of California Press.

Wallman, S. (1965). The communication of measurement in Basutoland, *Human Organization,* **24** (3), 236–43.

Wharton, C. H. (Editor) (1966). *Subsistence Agriculture and Economic Development.* Chicago: Aldine Publishing Co.

WHO (World Health Organization) (1985). *Women, Health and Development*

(Offset Publication no. 90). Geneva: WHO.

Whyte, W. F. (1981). *Participatory Approaches to Agricultural Research and Development: A state of the art paper.* Ithaca, New York: Rural Development Committee, Center for International Studies, Cornell University.

Willett, Anthony B. J. (1981). *Agricultural Group Development in Botswana.* 4 Vols. Gabarone: Government Printer.

Women and Food Information Network Newsletter (University of Arizona, Tucson; Office of International Agriculture) (1984). **2** (1).

World Bank (1975). *The Assault on World Poverty.* Baltimore: The Johns Hopkins University Press.

World Bank (1983a). *World Development Report.* Washington, DC: The World Bank.

World Bank (1983b). Tanzania Agricultural Sector Report. Washington, DC: The World Bank (Report no. 4052-TA).

World Bank (1983c). *Strengthening Research and Extension: the World Bank Experience* (Report no. 4684). Washington, DC: The World Bank.

Young, Michael, Perraton, Hilary, Jenkins, Janet and Dodds, Tony (1980). *Distance Teaching and the Third World. The Lion and the Clockwork Mouse.* London: Routledge and Kegan Paul.

Zandstra, H. G., Price, E. C., Litsinger, J. A. and Morris, R. A. (1981). *A Methodology for On-Farm Cropping Systems Research.* Los Baños, The Philippines: International Rice Research Institute.

Zeitlyn, Jonathan (1983). *Low Cost Printing for Development.* 4 Vols. New Delhi: CENDIT; and London: J. Zeitlyn.

Index

291